OIL AND GAS EXPLORATION AND PRODUCTION

OIL AND GAS EXPLORATION AND PRODUCTION

Deepak Sharma

RANDOM PUBLICATIONS

NEW DELHI - 110 002 (INDIA)

Oil and Gas Exploration and Production

ISBN 978-93-51116-67-7

Published in 2015 in India by

RANDOM PUBLICATIONS

4376-A/4B, Gali Murari Lal, Ansari Road
New Delhi-110 002
Phone: +9111-43580356, 23289044
E-mail: randomexports@gmail.com; sales@randompublications.com; info@randompublications.com

Reprinted 2018

Type Setting by: Friends Media, Delhi-110089
Digitally Printed at: Replika Press Pvt. Ltd.

REPRINT-2025

Preface

Hydrocarbon exploration is the search by petroleum geologists and geophysicists for hydrocarbon deposits beneath the Earth's surface, such as oil and natural gas. Oil and gas exploration are grouped under the science of petroleum geology. Exploration and production (E&P) is known as the upstream oil and gas industry. As global demand for energy soars, oil and gas fields are getting more and more complex to produce. Companies involved in the high-risk/high-reward area of exploration and production focus on finding, augmenting, producing and merchandising different types of oil and gas.

Oil and gas reserves are defined as volumes that will be commercially recovered in the future. Reserves are separated into three categories: proved, probable, and possible. To be included in any reserves category, all commercial aspects must have been addressed, which includes government consent. Technical issues alone separate proved from unproved categories. All reserve estimates involve some degree of uncertainty. The steps that lead to the production of oil and gas are diverse, complex and costly. They are diverse because the detection of oil and gas involves input from many specialties, ranging from geology to reservoir engineering. Visible surface features such as oil seeps, natural gas seeps, pockmarks provide basic evidence of hydrocarbon generation. However, most exploration depends on highly sophisticated technology to detect and determine the extent of these deposits using exploration geophysics. Areas thought to contain hydrocarbons are initially subjected to a gravity survey, magnetic survey, passive seismic or regional seismic reflection surveys to detect large-scale features of the sub-surface geology. Features of interest are subjected to more detailed seismic surveys which work on the principle of the time it takes for reflected sound waves to travel through matter (rock) of varying densities and using the process of depth conversion to create a profile of the substructure. Finally, when a prospect has been identified and evaluated and passes the oil company's selection criteria, an exploration well is drilled in an attempt to conclusively determine the presence or absence of oil or gas. Oil exploration is an expensive, high-risk operation. Offshore and remote area exploration is generally only undertaken by very large corporations or national governments. A prospect is a potential trap which geologists believe may contain hydrocarbons. A significant amount of geological, structural and seismic investigation must first be completed to redefine the potential hydrocarbon drill location from a lead to a prospect. Five geological factors have to be present for a prospect to work and if any of them fail neither oil nor gas will be present. In most nations the government issues licences to explore, develop and produce its oil and gas resources, which are typically administered by the oil ministry. There are several different types of licence. Oil companies often operate in joint ventures to spread the risk; one of the companies in the partnership is designated the operator who actually supervises the work.

This is an important book for engineering students and all those related to oil and gas industry.

I thank all members of my team who have helped in the preparation of the book. My special thanks go to "Random Publications" who have published the book.

— Deepak Sharma

Contents

Chapter 1

Extraction of Petroleum

The extraction of petroleum is the process by which usable petroleum is extracted and removed from the earth.

Figure: *Pumpjack on an oil well*

Locating the Oil Field

Geologists use seismic surveys to search for geological structures that may form oil reservoirs. The "classic" method includes making an underground explosion nearby and observing the seismic response that provides information about the geological structures under the ground 1. However, "passive" methods that extract information from naturally-occurring seismic waves are also known.

Other instruments such as gravimetres and magnetometres are also sometimes used in the search for petroleum. Extracting crude oil normally starts with drilling wells into the underground reservoir. When an oil well has been tapped, a geologist (known on the rig as the "mudlogger") will note its presence. Such a "mudlogger" is known to be sitting on the rig. Historically, in the USA, some oil fields existed where the oil rose naturally to the surface, but most of these fields have long since been used up, except in certain places in Alaska. Often many wells (called *multilateral wells*) are drilled into the same reservoir, to ensure that the extraction rate will be economically viable. Also, some wells (*secondary wells*) may be used to pump water, steam, acids or various gas mixtures into the reservoir to raise or maintain the reservoir pressure, and so maintain an economic extraction rate.

Drilling

The oil well is created by drilling a hole into the earth with an oil rig. A steel pipe (casing) is placed in the hole, to provide structural integrity to the newly drilled wellbore. Holes are then made in the base of the well to enable oil to pass into the bore. Finally a collection of valves called a "Christmas Tree" is fitted to the top, the valves regulating pressures and controlling flows.

Oil Extraction and Recovery

Primary Recovery

During the *primary recovery stage*, reservoir drive comes from a number of natural mechanisms. These include: natural water displacing oil downward into the well, expansion of the natural gas at the top of the reservoir, expansion of gas initially dissolved in the crude oil, and gravity drainage resulting from the movement of oil within the reservoir from the upper to the lower parts where the wells are located. Recovery factor during the primary recovery stage is typically 5-15%.

While the underground pressure in the oil reservoir is sufficient to force the oil to the surface, all that is necessary is to place a complex arrangement of valves (the Christmas tree) on the well head to connect the well to a pipeline network for storage and processing.

Secondary Recovery

Over the lifetime of the well the pressure will fall, and at some point there will be insufficient underground pressure to force the oil to the surface. After natural reservoir drive diminishes, *secondary recovery* methods are applied. They rely on the supply of external energy into the reservoir in the form of injecting fluids to increase reservoir pressure, hence replacing or increasing the natural reservoir drive with an artificial drive. Sometimes pumps, such as beam pumps and electrical submersible pumps (ESPs), are used to bring the oil to the surface. Other secondary recovery techniques increase the reservoir's pressure by water injection, natural gas reinjection and gas lift, which injects air, carbon dioxide or some other gas into the bottom of an active well, reducing the overall density of fluid in the wellbore. Typical recovery factor from water-flood operations is about 30%, depending on the properties of oil and the characteristics of the reservoir rock. On average, the recovery factor after primary and secondary oil recovery operations is between 35 and 45%.

Tertiary Recovery

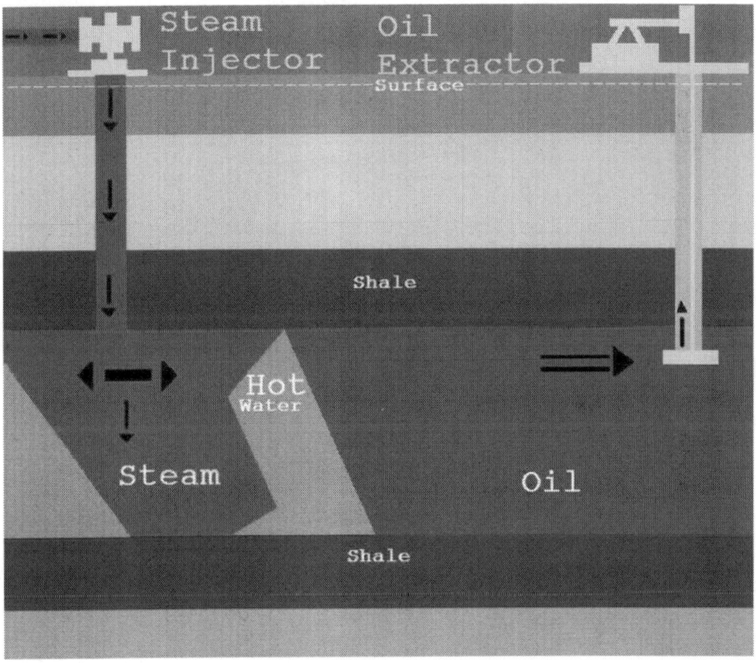

Figure: *Steam is injected into many oil fields where the oil is thicker and heavier than normal crude oil*

Tertiary, or enhanced oil recovery methods increase the mobility of the oil in order to increase extraction. Thermally enhanced oil recovery methods (TEOR) are tertiary recovery techniques that heat the oil, thus reducing its viscosity and making it easier to extract. Steam injection is the most common form of TEOR, and is often done with a cogeneration plant. In this type of cogeneration plant, a gas turbine is used to generate electricity and the waste heat is used to produce steam, which is then injected into the reservoir. This form of recovery is used extensively to increase oil extraction in the San Joaquin Valley, which has very heavy oil, yet accounts for 10% of the United States' oil extraction. In-situ burning is another form of TEOR, but instead of steam, some of the oil is burned to heat the surrounding oil.

Occasionally, surfactants (detergents) are injected to alter the surface tension between the water and oil in the reservoir, mobilising oil which would otherwise remain in the reservoir as residual oil.

Another method to reduce viscosity is carbon dioxide flooding. Tertiary recovery allows another 5% to 15% of the reservoir's oil to be recovered. Tertiary recovery begins when secondary oil recovery isn't enough to continue adequate extraction, but only when the oil can still be extracted profitably. This depends on the cost of the extraction method and the current price of crude oil. When prices are high, previously unprofitable wells are brought back into use and when they are low, extraction is curtailed.

Microbial treatments is another tertiary recovery method. Special blends of the microbes are used to treat and break down the hydrocarbon chain in oil thus making the oil easy to recover as well as being more economic versus other conventional methods. In some states, such as Texas, there are tax incentives for using these microbes in what is called a secondary tertiary recovery. Very few companies supply these, however companies like Bio Tech, Inc. have proven very successful in waterfloods across Texas.

Recovery Rates and Factors

The amount of oil that is recoverable is determined by a number of factors including the permeability of the rocks, the strength of natural drives (the gas present, pressure from adjacent water or gravity), and the viscosity of the oil. When the reservoir rocks are "tight" such as shale, oil generally cannot flow through but when they are permeable such as in sandstone, oil flows freely. The flow of oil is often helped by natural pressures surrounding the reservoir rocks including natural gas that may be dissolved in the oil, natural gas present above the oil, water below the oil and the strength of gravity. Oils tend to span a large range of viscosity from liquids as light as gasoline to heavy as tar. The lightest forms tend to result in higher extraction rates. Petroleum engineering is the discipline responsible for evaluating which well locations and recovery mechanisms are appropriate for a reservoir and for estimating recovery rates and oil reserves prior to actual extraction.

Estimated Ultimate Recovery

Although *ultimate recovery* of a well cannot be known with certainty until the well ceases production, petroleum engineers will often estimate an estimated ultimate recovery (EUR) based on decline rate projections years into the future. Various models, mathematical techniques and approximations are used.

Shale gas companies may intentionally choose methods that tend to extend the estimated production life of the well, or use models that estimate slower rates of production decline, in order to boost EURs, estimated returns, and stock prices.

Hydraulic Fracturing

Hydraulic fracturing is the propagation of fractures in a rock layer caused by the presence of a pressurised fluid. Hydraulic fractures form naturally, as in the case of veins or dikes, and is one means by which gas and petroleum from source rocks may migrate to reservoir rocks.

Figure: *BarnettShaleDrilling-9323*

Energy companies may attempt to accelerate this process in order to release petroleum, natural gas, coal seam gas, or other substances for extraction, via a technique called induced hydraulic fracturing (illustration), often shortened to fracking or hydrofracking. This type of fracturing, known colloquially as a 'frac job', creates fractures from a wellbore drilled into reservoir rock formations. A distinction can be made between low-volume hydraulic fracturing used to stimulate high-permeability reservoirs, which may consume typically 20,000 to 80,000 gallons of fluid per well, with high-volume hydraulic fracturing, used in the completion of tight gas and shale gas wells; high-volume hydraulic fracturing can use as much as two to three million gallons of fluid per well.

This latter practice has come under scrutiny internationally, with some countries suspending it, or even banning it completely. The first frac job was performed in 1947, though the current fracking technique was first used in the late 1990s in the Barnett Shale in Texas. The energy from the injection of a highly-pressurised fracking fluid creates new channels in the rock which can increase the extraction rates and ultimate recovery of fossil fuels.

Globally, the expectation that gas use will rise by more than 50% compared to 2010 levels, and account for over 25% of world energy demand in 2035 has led the International Energy Agency in 2011 to herald a "golden age of gas". Proponents of fracking point to the vast amounts of formerly in accessible hydrocarbons the process can extract. However, there remain large uncertainties in the amount of gas reserves that can be accessed in this way. Some have said that hydraulic fracturing would give the United States energy independence, but others disagree. Oil and gas companies have sought and received permission from the US Department of Energy (DOE) to export oil and gas to Europe and Asia (e.g., China and Korea) via new pipelines to export terminals on the Gulf Coast and East Coast. Higher prices abroad have also led to acquisition of US natural gas companies and shares of companies and their shale plays by Chinese, Japanese, and French companies

There are environmental concerns regarding the contamination of ground water, risks to air quality, the migration of gases and hydraulic fracturing chemicals to the surface, Complaints from a few residents on water quality in a developed natural gas field prompted an EPA groundwater investigation in Wyoming. The EPA reported detections of methane and other chemicals such as phthalates in private water wells and the health effects of these. Some are concerned that lower gas prices have also reduced the incentive to investigate renewable sources of energy, perpetuating dependence on gas and oil.

Mechanics

Fracturing in rocks at depth is suppressed by the confining pressure, due to the load caused by the overlying rock strata. This is particularly so in the case of 'tensile' (Mode 1) fractures, which require the walls of the fracture to move apart, working against this confining pressure. Hydraulic fracturing occurs when the effective stress is reduced sufficiently by an increase in the pressure of fluids within the rock, such that the minimum principal stress becomes tensile and exceeds the tensile strength of the material. Fractures formed in this way will typically be oriented perpendicularly to the minimum principal stress and for this reason, induced hydraulic fractures in wellbores are sometimes used to determine stress orientations. In natural examples, such as dikes or vein-filled fractures, their orientations can be used to infer past stress states.

Natural Examples

Rocks often contain evidence of past hydraulic fracturing events.

Veins

Most vein systems are a result of repeated hydraulic fracturing during periods of relatively high pore fluid pressure. This is particularly clear in the case of 'crack-seal' veins, where the vein material can be seen to have been added in a series of discrete fracturing events, with extra vein material deposited on each occasion. One mechanism to explain such examples of long-lasting repeated fracturing, is the effects of seismic activity, in which the stress levels rise and fall episodically and large volumes of fluid may be expelled from fluid-filled fractures during earthquakes, a process referred to as 'seismic pumping'.

Dikes

High-level minor intrusions such as dikes propagate through the crust in the form of fluid-filled cracks, although in this case the fluid is magma. In sedimentary rocks with a significant water content the fluid at the propagating fracture tip will be steam.

Induced Hydraulic Fracturing

The technique of hydraulic fracturing is used to increase or restore the rate at which fluids, such as oil, water, or natural gas can be produced from subterranean natural reservoirs. Reservoirs are typically porous sandstones, limestones or dolomite rocks, but also include 'unconventional reservoirs' such as shale rock or coal beds. Hydraulic fracturing enables the production of natural gas and oil from rock formations deep below the earth's surface (generally 5,000–20,000 feet or 1,500–6,100 m). At such depth, there may not be sufficient porosity, permeability or reservoir pressure to allow natural gas and oil to flow from the rock into the wellbore at economic rates. Thus, creating conductive fractures in the rock is essential to extract gas from shale reservoirs because of the extremely low natural permeability of shale, which is measured in the microdarcy to nanodarcy range. Fractures provide a conductive path connecting a larger area of the reservoir to the well, thereby increasing the area from which natural gas and liquids can be recovered from the targeted formation. So-called 'super fracking', which creates deeper cracks to release more oil and gas, will allow companies to frack more efficiently. The yield for a typical shale gas well generally falls off sharply after the first year or two.

While the main industrial use of hydraulic fracturing is in stimulating production from oil and gas wells, hydraulic fracturing is also applied to:

- Stimulating groundwater wells
- Preconditioning rock for caving or inducing rock to cave in mining
- As a means of enhancing waste remediation processes, usually hydrocarbon waste or spills
- Dispose of waste by injection into deep rock formations
- As a method to measure the stress in the earth
- For heat extraction to produce electricity in an enhanced geothermal systems History

The first frac job was performed in 1947 in limestone deposits. Significant R&D and technology demonstration were necessary before hydraulic fracturing could be commercially applied to shale gas deposits, due to shale's high porosity and low permeability. In the 1970s the federal government initiated both the Eastern Gas Shales Project, a set of dozens of public-private hydro-fracturing pilot demonstration projects, and the Gas Research Institute, a gas industry research consortium that received approval for research and funding from the Federal Energy Regulatory Commission.

Over this time, Sandia National Laboratories was conducting research into microseismic imaging for use in coalbeds, a geologic mapping technique that would prove crucial for the commercial recovery of natural gas from shale as well as oil from offshore drilling rigs. In the late 1970s, the Department of Energy pioneered massive hydraulic fracturing (MHF), a drilling technique that would be improved upon for the economic recovery of shale gas in the future. In 1986, a joint DOE-private venture completed the first successful multi-fracture horizontal well in shale. The Department of Energy later subsidised Mitchell Energy's first successful horizontal drill in the north-Texas Barnett Shale in 1991. Mitchell Energy engineers would go on to develop the hydraulic fracturing technique known as 'slickwater fracturing' that started the modern shale gas boom.

Method

A hydraulic fracture is formed by pumping the fracturing fluid into the wellbore at a rate sufficient to increase pressure downhole to exceed that of the fracture gradient of the rock. The rock cracks and the fracture fluid continues farther into the rock, extending the crack still farther, and so on. Operators typically try to maintain "fracture width", or slow its decline, following treatment by introducing a proppant into the injected fluid, a material, such as grains of sand, ceramic, or other particulates, that prevent the fractures from closing when the injection is stopped. Consideration of proppant strengths and prevention of proppant failure becomes more important at deeper depths where pressure and stresses on fractures are higher. The propped fracture is permeable enough to allow the flow of formation fluids to the well. Formation fluids include gas, oil, salt water, fresh water and fluids introduced to the formation during completion of the well during fracturing.

The location of one or more fractures along the length of the borehole is strictly controlled by various different methods which create or seal-off holes in the side of the wellbore. Typically, hydraulic fracturing is performed in cased wellbores and the zones to be fractured are accessed by perforating the casing at those locations.

Well Types

While hydraulic fracturing is many times performed in vertical wells, today it is also performed in horizontal wells. When done in already highly-permeable reservoirs such as sandstone-based wells, the technique is known as "well stimulation".

Horizontal drilling involves wellbores where the terminal drillhole is completed as a 'lateral' that extends parallel with the rock layer containing the substance to be extracted. For example, laterals extend 1,500 to 5,000 feet in the Barnett Shale basin in Texas, and up to 10,000 feet in the Bakken formation in North Dakota. In contrast, a vertical well only accesses the thickness of the rock layer, typically 50–300 feet. Horizontal drilling also reduces surface disruptions as fewer wells are required. Drilling usually induces damage to the pore space at the wellbore wall, reducing the permeability at and near the wellbore. This reduces flow into the borehole from the surrounding rock formation, and partially seals off the borehole from the surrounding rock. Hydraulic fracturing can be used to restore permeability.

Hydraulic fracturing is commonly applied to wells drilled in low permeability reservoir rock.

Fracturing

The fluid injected into the rock is typically a slurry of water, proppants, and chemical additives. Additionally, gels, foams, and compressed gases, including nitrogen, carbon dioxide and air can be injected. Various types of proppant include silica sand, resin-coated sand, and man-made ceramics. These vary depending on the type of permeability or grain strength needed. Sand containing naturally radioactive minerals is sometimes used so that the fracture trace along the wellbore can be measured. Chemical additives are applied to tailor the injected material to the specific geological situation, protect the well, and improve its operation, though the injected fluid is approximately 98-99.5% percent water, varying slightly based on the type of well. The composition of injected fluid is sometimes changed as the fracturing job proceeds. Often, acid is initially used to scour the perforations and clean up the near-wellbore area. Afterward, high pressure fracture fluid is injected into the wellbore, with the pressure above the fracture gradient of the rock. This fracture fluid contains water-soluble gelling agents (such as guar gum) which increase viscosity and efficiently deliver the proppant into the formation. As the fracturing process proceeds, viscosity reducing agents such as oxidisers and enzyme breakers are sometimes then added to the fracturing fluid to deactivate the gelling agents and encourage flowback. The proppant's purpose is primarily to provide a permeable and permanent filler to fill the void created during the fracturing process. At the end of the job the well is commonly flushed with water (sometimes blended with a friction reducing chemical) under pressure. Injected fluid is to some degree recovered and is managed by several methods, such as underground injection control, treatment and discharge, recycling, or temporary storage in pits or containers while new technology is being developed to better handle wastewater and improve reusability. Although the concentrations of the chemical additives are very low, the recovered fluid may be harmful due in part to hydrocarbons picked up from the formation.

Hydraulic fracturing equipment used in oil and natural gas fields usually consists of a slurry blender, one or more high pressure, high volume fracturing pumps (typically powerful triplex, or quintiplex pumps) and a monitoring unit. Associated equipment includes fracturing tanks, one or more units for storage and handling of proppant, high pressure treating iron, a chemical additive unit (used to accurately monitor chemical addition), low pressure flexible hoses, and many gauges and metres for flow rate, fluid density, and treating pressure. Fracturing equipment operates over a range of pressures and injection rates, and can reach up to 100 MPa (15,000 psi) and 265 L/s (100 barrels per minute).

Fracture Monitoring

Injection of radioactive tracers along with the other substances in hydraulic fracturing fluid is used to determine the injection profile and location of fractures created by hydraulic fracturing. US Patent No. 5635712 (George L. Scott III, Halliburton Company, 03-June-1997) describes this process. The patent lists a large number of gamma-emitting tracer isotopes that can be used as radioactive tracer material, including Gold-198,Xenon-133, Iodine-131, Rubidium-86, Chromium-51, Iron-59, Antimony-124, Strontium-85, Cobalt-58, Iridium-192, Scandium-46, Zinc-65, Silver-110, Cobalt-57, Cobalt-60, and Krypton-85. A 1983 patent, by Walter H. Fertl, for Dresser Industries (Dallas, Texas) also lists Iodine-131 as a potential tracer, along with Scandium-46, Zirconium-95, and Iridium-192. A 1995 patent by George L. Scott III, this time for The Energex Company (patent no. US5441110), also lists Iodine-131 as a suitable gamma-emitting tracer isotope, along with Potassium-39 (activated to Potassium-40), Potassium-41 (activated to Potassium-42), Potassium-43, Scandium-45 (activated to Scandium-46), Scandium-47, Scandium-48, Iodine-127, Iodine-128, Iodine-129, Iodine-130, Antimony-121, Antimony-122, Antimony-123, Antimony-124, Antimony-125, Antimony-126, and Antimony-127.

Measurements of the pressure and rate during the growth of a hydraulic fracture, as well as knowing the properties of the fluid and proppant being injected into the well provides the most common and simplest method of monitoring a hydraulic fracture treatment. This data, along with knowledge of the underground geology can be used to model information such as length, width and conductivity of a propped fracture.

For more advanced applications, Microseismic monitoring is sometimes used to estimate the size and orientation of hydraulically induced fractures. Microseismic activity is measured by placing an array of geophones in a nearby wellbore. By mapping the location of any small seismic events associated with the growing hydraulic fracture, the approximate geometry of the fracture is inferred. Tiltmetre arrays, deployed on the surface or down a well, provide another technology for monitoring the strains produced by hydraulic fracturing.

Emission of gases displaced by hydraulic fracturing into the atmosphere may be detected via atmospheric gas monitoring, and can be quantified directly via the eddy covariance flux measurements.

Horizontal Completions

Since the early 2000s, advances in drilling and completion technology has made drilling horizontal wellbores much more economical. Horizontal wellbores allow for far greater exposure to a formation than a conventional vertical wellbore. This is particularly useful in shale oil and gas formations which do not have sufficient permeability to produce economically with a vertical well. Such wells when drilled onshore are now usually hydraulically fractured many times, especially in North America. The type of wellbore completion used will affect how many times the formation is fractured, and at what locations along the horizontal section of the wellbore.

In North America, tight reservoirs such as the Bakken, Barnett Shale, Montney, Haynesville Shale and most recently Marcellus Shale are drilled, completed and fractured using this method. The method by which the fractures are placed along the wellbore is most commonly achieved by one of two methods, known as 'plug and perf' and 'sliding sleeve'.

The wellbore for a plug and perf job is generally composed of standard joints of steel casing, either cemented or uncemented, which is set in place at the conclusion of the drilling process. Once the drilling rig has been removed, a wireline truck is used to perforate near the end of the well, following which a fracturing job is pumped (commonly called a stage). Once the stage is finished, the wireline truck will set a plug in the well to temporarily seal off that section, and then perforate the next section of the wellbore. Another stage is then pumped, and the process is repeated as necessary along the entire length of the horizontal part of the wellbore.

The wellbore for the sliding sleeve technique is different in that the sliding sleeves are included at set spacings in the steel casing at the time it is set in place. The sliding sleeves are usually all closed at this time. When the well is ready to be fractured, using one of several activation techniques, the bottom sliding sleeve is opened and the first stage gets pumped. Once finished, the next sleeve is opened which concurrently isolates the first stage, and the process repeats. For the sliding sleeve method, wireline is usually not required.

These completion techniques may allow for more than 30 stages to be pumped into the horizontal section of a single well if required, which is far more than would typically be pumped into a vertical well.

Chemicals

Water is by far the largest component of fracking fluids. The initial drilling operation itself may consume from 65,000 gallons to 600,000 gallons of fracking fluids. Over its lifetime an average well will require up to an additional 5 million gallons of water for the initial fracking operation and possible restimulation frac jobs. The large volumes of water required have raised concerns about fracking in arid areas, such as Karoo in South Africa.

Chemical additives used in fracturing fluids typically make up less than 2% by weight of the total fluid. Over the life of a typical well, this may amount to 100,000 gallons of chemical additives. These additives (listed in a U.S. House of Representatives Report) include biocides, surfactants, viscosity-modifiers, and emulsifiers. They vary widely in toxicity: Many are used in household products such as cosmetics, lotions, soaps, detergents, furniture polishes, floor waxes, and paints, and some are used in food products. Some, however, are known carcinogens, some are toxic, and some are neurotoxins. For example: benzene (causes cancer, bone marrow failure), lead (damages the nervous system and causes brain disorders), ethylene glycol (antifreeze, causes death), methanol (highly toxic), boric acid (kidney damage, death), 2-butoxyethanol (causes hemolysis). Gamma-emitting isotopes (radioactive; can cause cancer) are also included in the fluid as tracers. Some of the isotopes used are Antimony-121, Antimony-122, Antimony-123, Antimony-124, Antimony-125, Antimony-126, Antimony-127, Chromium-51, Cobalt-57, Cobalt-58, Cobalt-60, Gold-198, Iodine-127, Iodine-128, Iodine-129, Iodine-130, Iodine-131, Iridium-192, Iron-59, Krypton-85, Lanthanum-140, Potassium-39 (activated to Potassium-40), Potassium-41 (activated to Potassium-42), Potassium-43, Rubidium-86, Scandium-45, Scandium-46, Scandium-47, Scandium-48, Silver-110, Strontium-85, Xenon-133, Zinc-65, and Zirconium-95. Several are typically combined and injected together. Their half lifes range from 40.2 hours (Lanthanum-140) to 5.27 years (Cobalt-60).

Despite concerns about the generally elevated radiation levels found near hydraulic fracturing sites and high levels of iodine-131 (a radioactive tracer used in hydraulic fracturing) found in drinking water and milk, iodine-131 is not listed among the chemicals to be monitored in the United States Environmental Protection Agency Hydraulic Fracturing Draft Study Plan. Other known radioactive tracers used in hydraulic fracturing but not listed as chemicals to be studied include radioactive isotopes of gold, xenon, rubidium, iridium, scandium, and krypton. Recently the EPA has not been very forthcoming regarding public disclosure of environmental contamination by the oil and gas industry.

The 2011 US House of Representatives investigative report on the chemicals used in hydraulic fracturing shows that of the 750 compounds in hydraulic fracturing products "more than 650 of these products contained chemicals that are known or possible human carcinogens, regulated under the Safe Drinking Water Act, or listed as hazardous air pollutants" (12).

The report also shows that between 2005 and 2009 279 products (93.6 million gallons-not including water) had at least one component listed as "proprietary" or "trade secret" on their Occupational Safety and Health Administration (OSHA) required Material Safety Data Sheet (MSDS). The MSDS is a list of chemical components in the products of chemical manufacturers, and according to OSHA, a manufacturer may withhold information designated as "proprietary" from this sheet.

When asked to reveal the proprietary components, most companies participating in the investigation were unable to do so, leading the committee to surmise these "companies are injecting fluids containing unknown chemicals about which they may have limited understanding of the potential risks posed to human health and the environment" (12). Without knowing the identity of the proprietary components, regulators cannot test for their presence. This prevents government regulators from establishing baseline levels of the substances prior to hydraulic fracturing and documenting changes in these levels, thereby making it impossible to prove that hydraulic fracturing is contaminating the environment with these substances. Third-party laboratories are performing analyses on soil, air, and water near the fracturing sites to measure the level of contamination by some of the known chemicals, but not the proprietary substances, involved in hydraulic fracturing. Each state has a contact person in charge of such regulation. A map of these contact people can be found at FracFocus.org as well.

Another 2011 study identified 632 chemicals used in natural gas operations. Only 353 of these are well-described in the scientific literature; and of these, more than 75% could affect skin, eyes, respiratory and gastrointestinal systems; roughly 40-50% could affect the brain and nervous, immune and cardiovascular systems and the kidneys; 37% could affect the endocrine system; and 25% were carcinogens and mutagens. The study indicated possible long-term health effects that might not appear immediately. The study recommended full disclosure of all products used, along with extensive air and water monitoring near natural gas operations; it also recommended that fracking's exemption from regulation under the US Safe Drinking Water Act be rescinded.

Some states have started requiring natural gas companies to "disclose the names of all chemicals to be stored and used a drilling site," keeping a record on file at the state's environmental agency, such as the case in Pennsylvania with the Department of Environmental Protection and in New York with the Department of Environmental Conservation. However, the continuing concern of some activists who oppose hydraulic fracturing is the lack of information really provided.

According to Weston Wilson in *Affirming Gasland*, "about 50% or so of these MSDS sheets lack a specific chemical name, and some MSDS sheets simply claim 'proprietary' status and list none of the chemicals in that container." As a result, some activists are calling for specific disclosure of chemicals used, such as the Chemical Abstract Service (CAS) number and specific chemical formulas, and increased access to such information. In his State of the Union address for 2012, Barack Obama stated his intention to force fracking companies to disclose the chemicals they use, though the subsequent, proposed guidelines were criticised for failing to specify how drillers will disclose the chemicals they use.

Terminology

Fracture Gradient

The pressure to fracture the formation at a particular depth divided by the depth. A fracture gradient of 18 kPa/m (0.8 psi/foot) implies that at a depth of 3 km (10,000 feet) a pressure of 54 MPa (8,000 psi) will extend a hydraulic fracture.

ISIP — Initial Shut In Pressure

The pressure measured immediately after injection stops. The ISIP provides a measure of the pressure in the fracture at the wellbore by removing contributions from fluid friction.

Leakoff

Loss of fracturing fluid from the fracture channel into the surrounding permeable rock.

Fracturing Fluid

The fluid used during a hydraulic fracture treatment of oil, gas, or water wells. The fracturing fluid has three major functions:

1. Open and extend the fracture.
2. Transport the proppant along the fracture length.
3. Transport radioactive tracers through the fractures to determine the injection profile and track the locations of fractures.

Proppant

Suspended particles in the fracturing fluid that are used to hold fractures open after a hydraulic fracturing treatment, thus producing a conductive pathway that fluids can easily flow along. Naturally occurring sand grains or artificial ceramic material are common proppants used.

Environmental Concerns

Environmental concerns with hydraulic fracturing include the potential contamination of ground water, risks to air quality, the potential migration of gases and hydraulic fracturing chemicals to the surface, the potential mishandling of waste, and the health effects of these, like cancer. Many cases of suspected groundwater contamination have been documented. Valerie Brown, a researcher and science writer, believes that with the explosive growth of natural gas wells in the US, public exposure to the many chemicals involved in hydraulic fracturing is increasing, with uncertain consequences.

Challenges to Research

It has been reported that Industry and governmental pressure have made it difficult to conduct and report the results of comprehensive studies of hydraulic fracturing. EPA investigations into the oil and gas industry's environmental impact have been narrowed in scope and/or had negative findings removed due to industry and government pressure A 2012 Cornell University report noted that it was difficult to assess health impact because of legislation, proprietary secrecy, and non-disclosure agreements that allow them to keep the proprietary chemicals used in the fluid secret. Cornell researcher Bamberger stated that if you don't know what chemicals are, you can't conduct pre-drilling tests and establish a baseline to prove that chemicals found postdrilling are from hydraulic fracturing. The researchers recommended requiring disclosure of all hydraulic fracturing fluids, that nondisclosure agreements not be allowed when public health is at risk, testing animals raised near hydraulic fracturing sites and animal products (milk, cheese, etc.) from animal raised near hydraulic fracturing sites prior to selling them to market, monitoring of water, soil and air more closely, and testing the air, water, soil and animals prior to drilling and at regular intervals thereafter. In addition, after court cases concerning contamination from hydraulic fracturing are settled, the documents are sealed. While the American Petroleum Institute deny that this practice has hidden problems with gas drilling, others believe it has and could lead to unnecessary risks to public safety and health.

The New York Times reported that the results of the 2004 EPA Study were censored due to strong industry influence and political pressure, however EPA agency officials stood by the results. An early draft of the study discussed the possibility of dangerous levels of fracking fluid contamination, and mentioned "possible evidence" of aquifer contamination. The final report concluded simply that fracking "poses little or no threat to drinking water". The study's scope was narrowed so that it only focused on the injection of fracking fluids, ignoring other aspects of the process such as disposal of fluids, and environmental concerns such as water quality, fish kills and acid burns. The study was concluded before public complaints of contamination started emerging. The study's conclusion that the injection of fracking fluids into coalbed methane (CBM) wells posed a minimal threat to underground drinking water sources may have influenced the 2005 Congressional decision to exempt hydraulic fracturing from regulation under the Safe Drinking Water Act.

The 2012 EPA Hydraulic Fracturing Draft Plan was also narrowed. It does not include studying the effects of iodine-131 (found in Philadelphia's drinking water) or other radioactive tracer isotopes used in hydraulic fracturing. Nor does the draft plan include evaluating the impact of wastewater. Christopher Portier, director of the US CDC's National Centre for Environmental Health and the Agency for Toxic Substances and Disease Registry, argued that, in addition to the EPA's plans to investigate the impact of fracking on drinking water, additional studies should be carried out to determine whether wastewater from the wells can harm people or animals and vegetables they eat. A group of US doctors called for a moratorium on fracking in populated areas until such studies had been done.

Proponents of hydraulic fracturing have erroneously reported in the press and other media that the recent University of Texas Study ("Fact-Based Regulation for Environmental Protection in Shale Gas Development") found that hydraulic fracturing caused no environmental contamination, when in fact the study found that all steps in the process except the actual injection of the fluid (which proponents artificially separated from the rest of the process and designated "hydraulic fracturing") have resulted in environmental contamination. The radioactivity of the injected fluid itself was not assessed in the University of Texas study. While the EPA recognises the potential for contamination of water by hydraulic fracturing, EPA Administrator Lisa Jackson testified in a Senate Hearing Committee stating that she is not aware of any proof where the fracking process itself has contaminated water.

Air Emissions and Pollution

One group of emissions associated with natural gas development and production, are the emissions associated with combustion. These emissions include particulate matter, nitrogen oxides, sulphur oxide, carbon dioxide and carbon monoxide. Another group of emissions that are routinely vented into the atmosphere are those linked with natural gas itself, which is composed of methane, ethane, liquid condensate, and volatile organic compounds (VOCs). The VOCs that are especially impactful on health are benzene, toluene, ethyl benzene, and xylene (referred to as a group, called BTEX). Health effects of exposure to these chemicals include neurological problems, birth defects, and cancer.

VOCs, including BTEX, mixed with nitrogen oxides from combustion and combined with sunlight can lead to ozone formation. Ozone has been shown to impact lung function, increase respiratory illness, and is particularly dangerous to lung development in children. In 2008, measured ambient concentrations in the rural Sublette County, Wyoming where ranching and natural gas are the main industries were frequently above the National Ambient Air Quality Standards (NAAQS) of 75ppb and have been recorded as high as 125 ppb. However, a study for the city of Fort Worth, Texas, examining air quality around natural gas sites "did not reveal any significant health threats." The Fort Worth Star-Telegram characterised that report as "the most comprehensive study of urban gas drilling to date."

Groundwater Contamination

Despite barriers to research, there are documented incidents of contamination. As early as 1987, an E.P.A. report was published that indicated fracture fluid invasion into James Parson's water well in Jackson County, West Virginia. The well, drilled by Kaiser Exploration and Mining Company, was found to have induced fractures that created a pathway to allow fracture fluid to contaminate the groundwater from which Mr. Parson's well was producing. The oil and gas industry and the E.P.A. disagreed regarding the accuracy and thoroughness of this report. In 2006 drilling fluids and methane were detected leaking from the ground near a gas well in Clark, Wyoming; 8 million cubic feet of methane were eventually released, and shallow groundwater was found to be contaminated. Directed by Congress, the U.S. EPA announced in March 2010 that it will examine claims of water pollution related to hydraulic fracturing.

In Garfield County, Colorado, another area with a high concentration of drilling rigs, volatile organic compound emissions increased 30%

between 2004 and 2006; during the same period there was a rash of health complaints from local residents. Epidemiological studies that might confirm or rule out any connection between these complaints and fracking are virtually non-existent. The health effects of VOCs are largely unquantified, so any causal relationship is difficult to ascertain; however, some of these chemicals are suspected carcinogens and neurotoxins. Investigators from the Colorado School of Public Health performed a study in Garfield regarding potential adverse health effects, and concluded that residents near gas wells might suffer chemical exposures, accidents from industry operations, and psychological impacts such as depression, anxiety and stress. This study (the only one of its kind to date) was never published, owing to disagreements between community members and the drilling company over the study's methods.

In 2009 13 water wells Dimock, Pennsylvania were contaminated with methane (one blew up). Cabot Oil & Gas had to financially compensate residents and construct a pipeline to bring in clean water. The company continues to deny that any "of the issues in Dimock have anything to do with hydraulic fracturing". The devices needed to prevent such water contamination cost as little as $600. Confusion remains regarding whether the water in Dimock is safe to drink. On Jan. 19, 2012, the EPA reversed its position, and asked that the agency's hazardous site cleanup division take immediate action to protect public health and safety.

In 2010 the film *Gasland* premiered at the Sundance Film Festival. The filmmaker claims that chemicals including toxins, known carcinogens, and heavy metals polluted the ground water near well sites in Pennsylvania, Wyoming, and Colorado.

Complaints about water quality from residents near a gas field in Pavillion, Wyoming prompted an EPA groundwater investigation. The EPA reported detections of methane and other chemicals such as phthalates in private water wells. An EPA draft report dated December 8, 2011 suggested that the ground water in the Pavillion, Wyoming, aquifer contains "compounds likely associated with gas production practices, including hydraulic fracturing". The EPA discovered traces of methane and foaming agents in several water wells near a gas rig. Samples of water taken from EPA's deep monitoring wells in the aquifer were found to contain gasoline, diesel fuel, BTEX (benzene, toluene, ethylbenzene, xylene), naphthalenes, isopropanol, and synthetic chemicals (e.g., glycols and alcohols) used in gas production and hydraulic fracturing fluid, and high methane levels. Benzene concentrations in the samples were well above Safe Drinking Water Act standards. The EPA report stated concerns about the movement of contaminants within the aquifer and the future safety of drinking water in the context of the area's complex geology. EPA's sampling of Pavillion area drinking water wells found chemicals consistent with those reported in previous EPA reports, including but not limited to methane and other petroleum hydrocarbons, indicating migration of contaminants from areas of gas production. The report also said that contaminants in wells near pits indicated that (frack) pits are a source of shallow ground water contamination. It also said, "Detections of organic chemicals are more numerous and exhibit higher concentrations in the deeper of the two monitoring wells ... (which) along with trends in methane, potassium, chloride, and pH, suggest a deep source of contamination." Their observations of chemical reactions in the field led them to suggest that upward migration of chemicals from deep underground is the culprit. They also found that the reports companies filed detailing jobs listed chemicals as a class or as "proprietary," "rendering identification of constituents impossible." The draft report also stated: "Alternative explanations were carefully con-sid-ered to explain individual sets of data. However, when considered together with other lines of evidence, the data indicates likely impact to ground water that can be explained by hydraulic fracturing." In response, in 2010 the U.S. Department of Health and Human Services' Agency for Toxic Substances and Disease Registry recommended that owners of tainted wells use alternate sources of water for drinking and cooking, and ventilation when showering. These recommendations were still in place as of December 2011. Encana is funding the alternate water supplies. During the investigation Luke Chavez (EPA investigator), commented that the contaminants could have come from cleaning products or oil and gas production, but said that in either case, their presence suggested problematic practices.

It is important to note that not every instance of groundwater methane contamination is a result of hydraulic fracturing. Often, local water wells drill through many shale and coal layers that can naturally seep methane into the producing groundwater. This methane is often biogenic (created by organic material decomposition) in origin as opposed to thermogenic (created through "thermal decomposition of buried organic material"). Thermogenic methane is the methane most often sought after by oil & gas companies deep in the earth, whereas biogenic methane is found in shallower formations (where water wells are typically drilled). Through isotope analysis and other detection methods, it is often fairly easy to determine whether the methane is biogenic or thermogenic, and thus determine from where it is produced. The presence of thermogenic methane does not confirm the source of gas. The gas composition and isotopic finger print must compared by experts with other known sources of gas to confirm a match.

In 2011 a Duke University study published in *Proceedings of the National Academy of Sciences* examined methane in groundwater in Pennsylvania and New York states overlying the Marcellus Shale and the Utica Shale. It determined that groundwater tended to contain much higher concentrations of methane near fracking wells, with potential explosion hazard; the methane's isotopic signatures and other geochemical indicators were consistent with it originating in the fracked deep shale formations, rather than any other source.

The 2011 University of Texas study described the environmental impact of each of the separate parts of the overall hydraulic fracturing process, or "phases of the shale gas development life cycle." These parts include of (1) drill pad construction and operation, (2) the construction, integrity, and performance of the wellbores, (3) the injection of the fluid once it is underground (which proponents consider the actual "fracking"), (4) the flowback of the fluid back towards the surface, (5) blowouts, often unreported, which spew hydraulic fracturing fluid and other byproducts across surrounding area, (5) integrity of other pipelines involved and (6) the disposal of the flowback, including waste water and other waste products. Associated problems include (1) Groundwater Contamination, (2) Blowouts and House Explosions, (3) Water Consumption and Supply, (4) Spill Management and Surface Water Protection, (5) Atmospheric Emissions, (6) Health Effects All but the injection stage were reported to be sources of contamination in the University of Texas study. The study concluded that if hydraulic fracturing is to be conducted in an environmentally safe manner, these issues need to be addressed first. Proponents have reported that groundwater contamination doesn't come directly from the "fracking" part of the process (the injection of hydraulic fracturing chemicals into Shale rock formations) but from other parts of the hydraulic fracturing process. Injection cannot be accomplished, however, without the accompanying stages. Poorly constructed or damaged wellbores and pipelines can allow the fluid to flow into aquifers. Volatile chemicals held in waste water evapouration ponds can to evapourate into the atmosphere, or overflow. In one of the cases described by a 2012 Cornell University study (conducted in Colorado, Louisiana, New York, Ohio, Pennsylvania and Texas) impounded wastewater was released into a field and pond, killing at least 70 animals. The runoff can also end up in groundwater systems. Groundwater may become contaminated by trucks carrying fracking chemicals and wastewater if they are involved in accidents on the way to fracking sites or disposal destinations. Disposal of fracking fluid by injection can cause earthquakes, and release of unprocessed or under-processed waste water into rivers can contaminate water supplies. Critics have noted that it is "difficult for researchers to be objective if their university receives a lot of grants and funds from the industry." An Energy Institute spokesperson said that the study was not funded by the industry.

He said funds came from the university, which has a variety of funding sources. There are extensive links between UT and the oil & gas industry, with the giving of Royal Dutch Shell to the university currently standing at more than $24.8 million, $4m alone having been handed over for 2012. Since 2011, Shell has partnered Texas in a program called Shell-UT Unconventional Research, and the university has a similar research program in place with Exxon Mobil. Halliburton, the largest supplier of fracking services in the United States, has also given millions of dollars to the university. Statoil announced a $5m research agreement (part of which will focus on oil shale) with UT's Bureau of Economic Geology in September 2011, whose program director, Ian Duncan, was the senior contributor for the parts of the Texas study to do with the environmental impacts of shale gas development.

In DISH, Texas, elevated levels of disulphides, benzene, xylenes and naphthalene have been detected in the air, alongside numerous local complaints of headaches, diarrhea, nosebleeds, dizziness, muscle spasms and other problems. Additionally, the Colorado Oil & Gas Conservation Commission has found some wells containing thermogenic methane due to oil and gas development upon investigating complaints from residents. Cause-and-effect relationships have not been established.

A 2011 report by the Massachusetts Institute of Technology addressed groundwater contamination, noting "There has been concern that these fractures can also penetrate shallow freshwater zones and contaminate them with fracturing ûuid, but there is no evidence that this is occurring. There is, however, evidence of natural gas migration into freshwater zones in some areas, most likely as a result of substandard well completion practices by a few operators. There are additional environmental challenges in the area of water management, particularly the effective disposal of fracture fluids". This study encourages the use of industry best practices to prevent such events from recurring.

Radioactive Contamination

The New York Times has reported radiation in hydraulic fracturing wastewater released into rivers in Pennsylvania. It collected data from more than 200 natural gas wells in Pennsylvania and has posted a map entitled Toxic Contamination from Natural Gas Wells in Pennsylvania. Sand containing gamma-emitting tracer isotopes is used to trace and measure fractures. The *Times* stated "never-reported studies" by the United States Environmental Protection Agency and a "confidential study by the drilling industry" concluded that radioactivity in drilling waste cannot be fully diluted in rivers and other waterways. Despite this, as of early 2011 federal and state regulators did not require sewage treatment plants that accept drilling waste (which is mostly water) to test for radioactivity. In Pennsylvania, where the drilling boom began in 2008, most drinking-water intake plants downstream from those sewage treatment plants have not tested for radioactivity since before 2006. *The New York Times* reporting has predictably been criticized by aggrieved parties, but one venerable science writer has taken issue with one instance of the newspaper's presentation and explanation of its calculations regarding dilution, charging that a lack of context made the article's analysis uninformative.

According to a *Times* report in February 2011, wastewater at 116 of 179 deep gas wells in Pennsylvania "contained high levels of radiation," but its effect on public drinking water supplies is unknown because water suppliers are required to conduct tests of radiation "only sporadically". The *New York Post* stated that the Pennsylvania Department of Environmental Protection reported that all samples it took from seven rivers in November and December 2010 "showed levels at or below the normal naturally occurring background levels of radioactivity", and "below the federal drinking water standard for Radium 226 and 228." However the samples taken by the state at least one river, (the Monongahela, a source of drinking water for parts of Pittsburgh), were taken upstream from the sewage treatment plants accepting drilling waste water.

In Pennsylvania, much of this wastewater from hydraulic fracturing operations is processed by public sewage treatment plants. However, many sewage plants say that they are incapable of removing the radioactive components of this waste, which is often released into major rivers. Industry officials, though, claim that these levels are diluted enough that public health is not compromised. This is a major concern as it provides the possibility for radioactive waste to enter into public water supplies. In April 2011, Environmental Protection

Agency (EPA) found elevated iodine-131 levels in Philadelphia's drinking water and milk from Little Rock, Arkansas. The National Cancer Institute has reported that children exposed to iodine-131, especially those drinking a great deal of milk, may have an increased risk of thyroid cancer. Both Philadelphia and Little Rock are located downstream from shale formations in which hydraulic fracturing is occurring. Iodine-131 was also found in the drinking water of other cities near other hydrofracturing sites.

In response to the Environmental Protection Agency (EPA) findings, the Philadelphia Water Department also posted a notice that Iodine-131 had been found in the water supply. The notice omits the fact that Iodine-131 is popular radioactive tracer used in hydrofracturing fluid to determine the injection profile and location of fractures created by hydraulic fracturing, and is mentioned in multiple hydraulic fracturing technology patents. In the notice, the Philadelphia Water Department attributes the presence of Iodine-131 to nuclear energy production and the March 2011 Japanese nuclear incident instead. When iodine-131 was still found in the Wissahickon Creek, and at several sewage treatment plants along the creek near Philadelphia in late July, long after the fallout from the Japanese incident would have decayed, Philadelphia Water Department officials became concerned. They reviewed Philadelphia's U.S. Environmental Protection Agency records and found that iodine-131 had been found in several Philadelphia drinking water samples long before the Fukushima accident. In fact, Environmental Protection Agency (EPA) records showed that Philadelphia's iodine-131 levels were the highest in the last decade in the set of those measured at 59 locations across the United States. The Philadelphia Water Department is currently attributing the elevated levels to cancer patients' urine from undetermined sources because iodine-131 is used in the treatment of thyroid cancer. The EPA and the Philadelphia Water Department are not specifically acknowledging that iodine-131, a radioactive tracer used in hydraulic fracturing since at least 1976, may be coming from the hydraulic fracturing wastewater released into the rivers. The EPA, however, is concerned about radionuclide levels in drinking water. The EPA has asked the Pennsylvania Department of Environmental Protection (PDEP) to require "community water systems (CWSs) near publicly owned treatment works (POTWs) and centralised wastewater treatment (CWT) facilities receiving Marcellus Shale wastewater to conduct sampling immediately for radionuclides." They note that "in previous monitoring, radionuclides were not detected or were detected at levels less than one-half of maximum contaminant levels," but that "the CWS have not sampled after the introduction of Marcellus Shale operations. The EPA letter adds that "Discharges from these operations could increase radionuclide levels substantially." The warning about iodine-131 was still posted on the Philadelphia Water Department web site as of February 25, 2012.

The New York Times has implicated the DEP in industry-friendly inactivity, requesting rather than requiring them to handle their own flowback waste rather than sending it to public water treatment facilities. However, former Pennsylvania DEP Secretary John Hanger, who served under Gov. Ed Rendell (D), has affirmed that municipal drinking water throughout the state is safe, but added that the environmentalists were accurate in stating that Pennsylvania's water treatment plants were not equipped to treat hydraulic fracturing water. Current Pennsylvania DEP Secretary Michael Krancer serving under Gov. Tom Corbett (R) has denied that untreated wastewater is being discharged into the state's waterways. It has been observed that Corbett received over a million dollars in gas industry contributions, more than all his competitors combined, during his election campaign *The New York Times* reported that regulations are lax in Pennsylvania.

The oil and gas industry is generally left to police itself in the case of accidents. Unannounced inspections are not made by regulators: the companies report their own spills, and create their own remediation plans. A recent review of the state-approved plans found them to appear to be in violation of the law. Treatment plants are still not equipped to remove radioactive material and are not required to test for it. Despite this, in 2009 the Ridgway Borough's public sewage treatment plant, in Elk County, PA, facility was sent wastewater containing radium and other types of radiation at at 275-780 times the drinking-water standard. The water being released from the plant was not tested for radiation levels. Part of the problem is that growth in waste produced by the industry has outpaced regulators and state resources. It should be noted that "safe drinking water standards" have not yet been set for many of the substances known to be in hydrofracturing fluids or their radioactivity levels, and their levels are not included in public drinking water quality reports.

Earthquakes

A report in the UK concluded that fracking was the likely cause of some small earth tremors that happened during shale gas drilling. In addition, the United States Geological Survey (USGS) reports that, "Earthquakes induced by human activity have been documented in a few locations" in the United States, Japan, and Canada, "the cause of which was injection of fluids into deep wells for waste disposal and secondary recovery of oil, and the use of reservoirs for water supplies." The disposal and injection wells referenced are regulated under the Safe Drinking Water Act and UIC laws and are not wells where hydraulic fracturing is generally performed.

Several earthquakes—including a substantial, magnitude 4.0 one on New Year's Eve—that had hit Youngstown, Ohio, throughout 2011 are likely linked to a disposal well for injecting wastewater used in the hydraulic fracturing process, according to seismologists at Columbia University.

Greenhouse Gas Emissions

The use of natural gas rather than oil or coal is often viewed as a way of alleviating global warming: natural gas burns more cleanly, and gas power stations can produce up to 50% less greenhouse gases than coal stations. However, an analysis by Howarth et al. of the well-to-consumer lifecycle of fracked natural gas concluded that 3.6–7.9% of the methane produced by a well will be leaked into the atmosphere during the well's lifetime. According to the analysis, methane is such a potent greenhouse gas, this means that over short timescales, shale gas is actually worse than coal or oil. Methane gradually breaks down in the atmosphere, forming carbon dioxide, so that over very long periods it is no more problematic than carbon dioxide; in the meantime, even if shale gas is burnt in efficient gas power stations, its greenhouse-gas footprint is still worse than coal or oil for timescales of less than fifty years. This analysis by Howarth et al. refers to the 2011 study by the same authors published in *Climatic Change Letters* in which they controversially claimed that the extraction of shale gas may lead to the emission of as much or more greenhouse gas emissions than oil or coal. However, several studies have argued that the paper was flawed and/or come to completely different conclusions, including assessments by experts at the US Department of Energy, by Carnegie Mellon University and the University of Maryland, as well as by the Natural Resources Defence Council, which concluded that the Howarth *et al.* paper's use of a 20-year time horizon for global warming potential of methane is "too short a period to be appropriate for policy analysis." In January 2012, Howarth's colleagues at Cornell University responded with their assessment, arguing that the Howarth paper was "seriously flawed" because it "significantly overestimates the fugitive emissions associated with unconventional gas extraction, undervalues the contribution of 'green technologies' to reducing those emissions to a level approaching that of conventional gas, bases their comparison between gas and coal on heat rather than electricity generation (almost the sole use of coal), and assumes a time interval over which to compute the relative climate impact of gas compared to coal that does not capture the contrast between the long residence time of CO2 and the short residence time of methane in the atmosphere."

The authors of that response conclude that "shale gas has a GHG footprint that is half and perhaps a third that of coal," based upon "more reasonable leakage rates and bases of comparison." Howarth et al. responded to this criticism: "We stand by our approach and findings. The latest EPA estimate for methane emissions from shale gas falls within the range of our estimates but not those of Cathles et al, which are substantially lower."

Public Relations

The considerable opposition against fracking activities in local townships has led companies to adopt a variety of public relations measures to assuage fears about fracking, including the admitted use of "military tactics to counter drilling oppo-nents". At a conference where public relations measures were discussed, a senior executive at Anadarko Petroleum was recorded on tape saying, while referring to fracking opponents. Matt Pitzarella, spokesman for the most important fracking company in Pennsylvania, Range Resources, also told other conference attendees that Range employed psychological warfare operations veterans. According to Pitzarella, the experience learnt in the Middle East has been valuable to Range Resources in Pennsylvania, when dealing with emotionally-charged township meetings and advising townships on zoning and local ordinances dealing with fracking. Furthermore, in a February 2012 campaign speech, Rick Santorum, a candidate for the 2012 Republican Party presidential nomination, claimed opponents of hydraulic fracturing were trying to scare people in a "reign of environmental terror". He accused President Obama of catering to "radical environmental groups" although President Obama supports the practice of hydraulic fracturing as well.

Hydraulic Fracturing by Country

Hydraulic fracturing has become a contentious environmental and health issue with France banning the practice and a moratorium in place in New South Wales (Australia), Karoo basin (South Africa), Quebec (Canada), and some of the states of the US.

Offshore Drilling

Offshore drilling refers to a mechanical process where a wellbore is drilled through the seabed. It is typically carried out in order to explore for and subsequently produce hydrocarbons which lie in rock formations beneath the seabed. Most commonly, the term is used to describe drilling activities on the continental shelf, though the term can also be applied to drilling in lakes, inshore waters and inland seas.

Offshore drilling presents environmental challenges, both from the produced hydrocarbons and the materials used during the drilling operation. Controversies include the ongoing US offshore drilling debate.

There are many different types of facilities from which offshore drilling operations take place. These include bottom founded drilling rigs (jackup barges and swamp barges), combined drilling and production facilities either bottom founded or floating platforms, and deepwater mobile offshore drilling units (MODU) including semi-submersibles and drillships. These are capable of operating in water depths up to 10,000 ft. In shallower waters the mobile units are anchored to the seabed, however in deeper water (>5,000 ft) the semisubmersibles or drillships are maintained at the required drilling location using dynamic positioning.

History

Around 1891, the first submerged oil wells were drilled from platforms built on piles in the fresh waters of the Grand Lake St. Marys (a.k.a. Mercer County Reservoir) in Ohio. The wells were developed by small local companies such as Bryson, Riley Oil, German-American, and Banker's Oil.

Around 1896, the first submerged oil wells in salt water were drilled in the portion of the Summerland field extending under the Santa Barbara Channel in California. The wells were drilled from piers extending from land out into the channel. Other notable early submerged drilling activities occurred on the Canadian side of Lake Erie in the 1900s and Caddo Lake in Louisiana in the 1910s. Shortly thereafter wells were drilled in tidal zones along the Texasand Louisiana gulf coast. The Goose Creek Oil Field near Baytown, Texas is one such example. In the 1920s drilling activities occurred from concrete platforms in Venezuela's Lake Maracaibo.

One of the oldest subsea wells is the Bibi Eibat well, which came on stream in 1923 in Azerbaijan. The well was located on an artificial island in a shallow portion of the Caspian Sea. In the early 1930s, the Texas Co., later Texaco (now Chevron) developed the first mobile steel barges for drilling in the brackish coastal areas of the Gulf of Mexico.

In 1937, Pure Oil (now Chevron) and its partner Superior Oil (now ExxonMobil) used a fixed platform to develop a field 1 mile offshore of Calcasieu Parish, Louisiana in 14 feet of water.

In 1946, Magnolia Petroleum (now ExxonMobil) drilled at a site 18 miles off the coast, erecting a platform in 18 feet of water off St. Mary Parish, Louisiana.

In early 1947, Superior Oil erected a drilling and production platform in 20 feet of water some 18 miles off Vermilion Parish, La. But it was Kerr-McGee Oil Industries (now Anadarko Petroleum), as operator for partners Phillips Petroleum (ConocoPhillips) and Stanolind Oil & Gas (BP) that completed its historic Ship Shoal Block 32 well in October 1947, months before Superior actually drilled a discovery from their Vermilion platform farther offshore. In any case, that made Kerr-McGee's well the first oil discovery drilled out of sight of land.

When offshore drilling moved into deeper waters of up to 100 feet, fixed platform rigs were built, until demands for drilling equipment was needed in the 100- to 400-foot depth of the Gulf of Mexico, the first jack-up rigsbegan appearing from specialised offshore drilling contractors such as forerunners of ENSCO International.

The first semi-submersible resulted from an unexpected observation in 1961. Blue Water Drilling Company owned and operated the four-column submersible Blue Water Rig No.1 in the Gulf of Mexico for Shell Oil Company. As the pontoons were not sufficiently buoyant to support the weight of the rig and its consumables, it was towed between locations at a draught mid-way between the top of the pontoons and the underside of the deck. It was noticed that the motions at this draught were very small, and Blue Water Drilling and Shell jointly decided to try operating the rig in the floating mode. The concept of an anchored, stable floating deep-sea platform had been designed and tested back in the 1920's by Edward Robert Armstrong for the purpose of operating aircraft with an invention known as the 'seadrome'. The first purpose-built drilling semi-submersible *Ocean Driller* was launched in 1963. Since then, many semi-submersibles have been purpose-designed for the drilling industry mobile offshore fleet.

The first offshore drillship was the *CUSS 1* developed for the Mohole project to drill into the Earth's crust.

As of June, 2010, there were over 620 mobile offshore drilling rigs (Jackups, semisubs, drillships, barges) available for service in the competitive rig fleet.

The world's deepest platform is currently the Perdido in the Gulf of Mexico, floating in 2,438 metres of water. It is operated by Royal Dutch Shell and was built at a cost of $3 billion.

Main Offshore Fields

Notable offshore fields today are found in the North Sea, the Gulf of Mexico, the Campos and Santos Basins off the coasts of Brazil, Newfoundland and Nova Scotia (Atlantic Canada), several fields off West Africa most notably west of Nigeria and Angola, as well as offshore fields in South East Asia and Sakhalin, Russia. Also major offshore oil fields are located in the Persian Gulf such as Safaniya, Manifa, and Marjan which belong to Saudi Arabia and are developed by Saudi Aramco.

Challenges

Offshore oil and gas production is more challenging than land-based installations due to the remote and harsher environment. Much of the innovation in the offshore petroleum sector concerns overcoming these challenges, including the need to provide very large production facilities. Production and drilling facilities may be very large and a large investment, such as the Troll A platform standing on a depth of 300 metres.

Another type of offshore platform may float with a mooring system to maintain it on location. While a floating system may be lower cost in deeper waters than a fixed platform, the dynamic nature of the platforms introduces many challenges for the drilling and production facilities. The ocean can add several hundred metres or more to the fluid column. The addition increases the equivalent circulating density and downhole pressures in drilling wells, as well as the energy needed to lift produced fluids for separation on the platform.

The trend today is to conduct more of the production operations subsea, by separating water from oil and re-injecting it rather than pumping it up to a platform, or by flowing to onshore, with no installations visible above the sea. Subsea installations help to exploit resources at progressively deeper waters, locations which have been inaccessible, and overcome challenges posed by sea ice, such as in the Barents Sea.

Offshore manned facilities also present logistics and human resources challenges. An offshore oil platform is a small community in itself with cafeteria, sleeping quarters, management, and other support functions. In the North Sea, staff members are transported by helicopter for a two-week shift. They usually receive higher salary than onshore workers do. Supplies and waste are transported by ship, and the supply deliveries need to be carefully planned because storage space on the platform is limited. Today, much effort goes into relocating as many of the personnel as possible onshore, where management and technical experts are in touch with the platform by video conferencing. An onshore job is also more attractive for the aging workforce in the petroleum industry, at least in the western world. These efforts among others are contained in the established term integrated operations. The increased use of subsea facilities helps achieve the objective of keeping more workers onshore. Subsea facilities are also easier to expand, with new separators or different modules for different oil types, and are not limited by the fixed floor space of an above-water installation.

Effects on the Environment

Offshore oil production involves environmental risks, most notably oil spills from oil tankers or pipelines transporting oil from the platform to onshore facilities, and from leaks and accidents on the platform. Produced water is also generated, which is excess water from oil or gas production and includes varying amounts of oil, or other chemicals used in, or resulting from, oil production. According to the organisation Culture Change, a Gulf of Mexico rig dumps about 90,000 tons of drilling fluid and cuttings over its lifetime, with its wells also contributing with heavy metals.

Chapter 2

Supply of Oil

Overall Supply Levels

According to the IEA's Oil Market Report dated 13 December 2011, global oil supply had risen to a record high of 90.0 mb/day by November 2011. Of this, oil supply from OPEC nations represented only 30.68 mb/day (34.1% of the total).

Discoveries

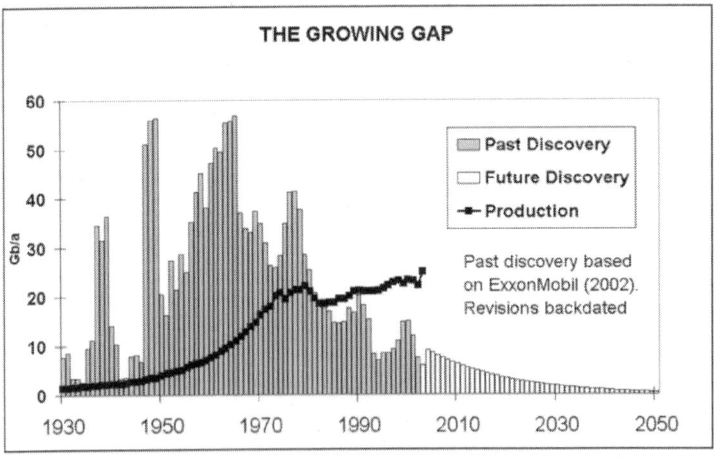

Figure : *Growing gap between discovery and production*

"All the easy oil and gas in the world has pretty much been found. Now comes the harder work in finding and producing oil from more challenging environments and work areas. " — William J. Cummings, Exxon-Mobil company spokesman, December 2005

"It is pretty clear that there is not much chance of finding any significant quantity of new cheap oil. Any new or unconventional oil is going to be expensive." — Lord Ron Oxburgh, a former chairman of Shell, October 2008

To pump oil, it first needs to be discovered. The peak of world oilfield discoveries occurred in 1965 at around 55 billion barrels (8.7×10^9 m^3)(Gb)/year. According to the Association for the Study of Peak Oil and Gas (ASPO), the rate of discovery has been falling steadily since. Less than 10 Gb/yr of oil were discovered each year between 2002-2007. According to a 2010 Reuters article, the annual rate of discovery of new fields has remained remarkably constant at 15-20 Gb/yr.

Reserves

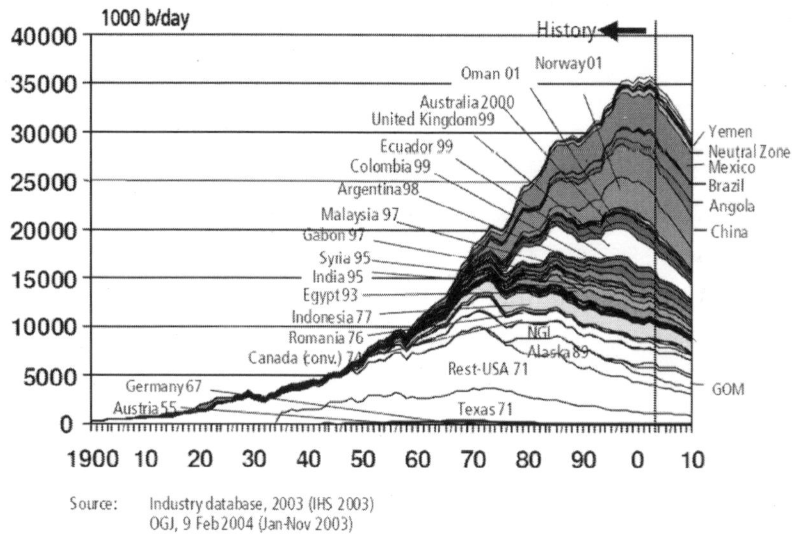

Figure : *2004 U.S. government predictions for oil production other than in OPEC and the former Soviet Union.*

Total possible conventional crude oil reserves include all crude oil with 90-95% certainty of being technically possible to produce (from reservoirs through a wellbore using primary, secondary, improved, enhanced, or tertiary methods), all crude with a 50% probability of being produced in the future, and discovered reserves which have a 5-10% possibility of being produced in the future. These are referred to as 1P/Proven (90-95%), 2P/Probable (50%), and 3P/Possible (5-10%). This does not include liquids extracted from mined solids or gasses (oil sands, oil shales, gas-to-liquid processes, or coal-to-liquid processes).

Many current 2P calculations predict reserves to be between 1150-1350 Gb, but because of misinformation, withheld information, and misleading reserve calculations, it has been reported that 2P reserves are likely nearer to 850-900 Gb. Reserves in effect peaked in 1980, when production first surpassed new discoveries, though creative methods of recalculating reserves have made this difficult to establish exactly.

Current technology is capable of extracting about 40% of the oil from most wells. Some speculate that future technology will make further extraction possible, but this future technology is usually already considered in Proven and Probable (2P) reserve numbers.

In many major producing countries, the majority of reserves claims have not been subject to outside audit or examination. Most of the easy-to-extract oil has been found. Recent price increases have led to oil exploration in areas where extraction is much more expensive, such as in extremely deep wells, extreme downhole temperatures, and environmentally sensitive areas or where high technology will be required to extract the oil. A lower rate of discoveries per explorations has led to a shortage of drilling rigs, increases in steel prices, and overall increases in costs due to complexity.

Concerns Over Stated Reserves

> *"World reserves are confused and in fact inflated. Many of the so-called reserves are in fact resources. They're not delineated, they're not accessible, they're not available for production."*
>
> *— Sadad I. Al-Husseini, former VP of Aramco, presentation to the Oil and Money conference, October 2007.*

Al-Husseini estimated that 300 billion barrels (48×10^9 m^3) of the world's 1,200 billion barrels (190×10^9 m^3) of proven reserves should be recategorised as speculative resources.

One difficulty in forecasting the date of peak oil is the opacity surrounding the oil reserves classified as 'proven'. Many worrying signs concerning the depletion of proven reserves have emerged in recent years. This was best exemplified by the 2004 scandal surrounding the 'evapouration' of 20% of Shell's reserves.

For the most part, proven reserves are stated by the oil companies, the producer states and the consumer states. All three have reasons to overstate their proven reserves: oil companies may look to increase their potential worth; producer countries gain a stronger international stature; and governments of consumer countries may seek a means to foster sentiments of security and stability within their economies and among consumers.

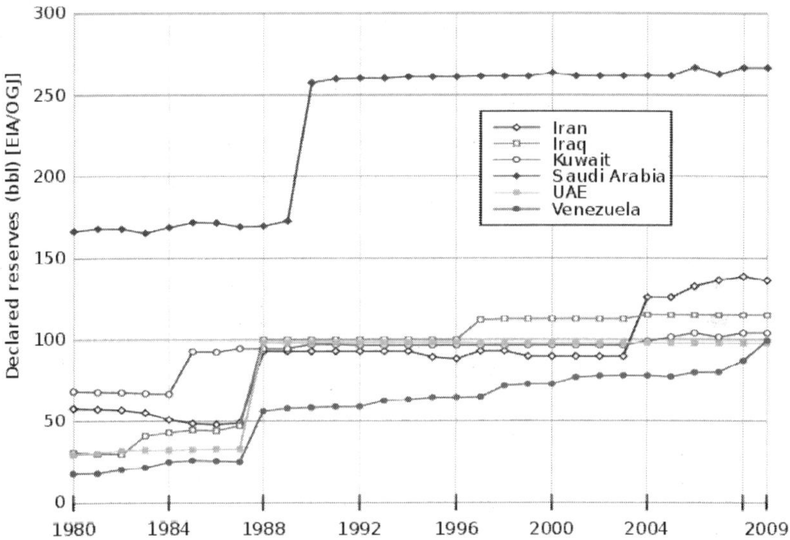

Figure : *Graph of OPEC reported reserves showing refutable jumps in stated reserves without associated discoveries, as well as the lack of depletion despite yearly production.*

Major discrepancies arise from accuracy issues with OPEC's self-reported numbers. Besides the possibility that these nations have overstated their reserves for political reasons (during periods of no substantial discoveries), over 70 nations also follow a practice of not reducing their reserves to account for yearly production. Analysts have suggested that OPEC member nations have economic incentives to exaggerate their reserves, as the OPEC quota system allows greater output for countries with greater reserves.

Kuwait, for example, was reported in the January 2006 issue of *Petroleum Intelligence Weekly* to have only 48 billion barrels (7.6×10^9 m^3) in reserve, of which only 24 were fully proven. This report was based on the leak of a confidential document from Kuwait and has not been formally denied by the Kuwaiti authorities. This leaked document is from 2001, so the figure includes oil that has been produced since 2001, roughly 5-6 billion barrels (950×10^6 m^3), but excludes revisions or discoveries made since then. Additionally, the reported 1.5 billion barrels (240×10^6 m^3) of oil burned off by Iraqi soldiers in the First Persian Gulf War are conspicuously missing from Kuwait's figures.

On the other hand, investigative journalist Greg Palast argues that oil companies have an interest in making oil look more rare than it is, to justify higher prices. This view is contested by ecological journalist Richard Heinberg. Other analysts argue that oil producing countries understate the extent of their reserves to drive up the price. In November 2009, a senior official at the IEA alleged that the United States had encouraged the international agency to manipulate depletion rates and future reserve data to maintain lower oil prices.

In 2005, the IEA predicted that 2030 production rates would reach 120,000,000 barrels per day (19,000,000 m^3/d), but this number was gradually reduced to 105,000,000 barrels per day (16,700,000 m^3/d). The IEA official alleged industry insiders agree that even 90 to 95,000,000 barrels per day (15,100,000 m^3/d) might be impossible to achieve. Although many outsiders had questioned the IEA numbers in the past, this was the first time an insider had raised the same concerns. A 2008 analysis of IEA predictions questioned several underlying assumptions and claimed that a 2030 production level of 75,000,000 barrels per day (11,900,000 m^3/d) (comprising 55,000,000 barrels (8,700,000 m^3) of crude oil and 20,000,000 barrels (3,200,000 m^3) of both non-conventional oil and natural gas liquids) was more realistic than the IEA numbers.

The EUR reported by the 2000 USGS survey of 2,300 billion barrels (370×10^9 m^3) has been criticized for assuming a discovery trend over the next twenty years that would reverse the observed trend of the past 40 years. Their 95% confidence EUR of 2,300 billion barrels (370×10^9 m^3) assumed that discovery levels would stay steady, despite the fact that discovery levels have been falling steadily since the 1960s. That trend of falling discoveries has continued in the ten years since the USGS made their assumption. The 2000 USGS is also criticized for introducing other methodological errors, as well as assuming 2030 production rates inconsistent with projected reserves.

Unconventional Sources

Unconventional sources, such as heavy crude oil, oil sands, and oil shale are not counted as part of oil reserves. However, with rule changes by the SEC, oil companies can now book them as proven reserves after opening a strip mine or thermal facility for extraction. These unconventional sources are more labour and resource intensive to produce, however, requiring extra energy to refine, resulting in higher production costs and up to three times more greenhouse gas emissions per barrel (or barrel equivalent) on a "well to tank" basis or 10 to 45% more on a "well to wheels" basis, which includes the carbon emitted from combustion of the final product. While the energy used, resources needed, and environmental effects of extracting unconventional sources has traditionally been prohibitively high, the three major unconventional oil sources being considered for large scale production are the extra heavy oil in the Orinoco Belt of Venezuela, the Athabasca Oil Sands in the Western Canadian Sedimentary Basin, and the oil shales of the Green River Formation in Colorado, Utah, and Wyoming in the United States. Energy companies such as Syncrude and Suncor have been extracting bitumen for decades but production has increased greatly in recent years with the development of Steam Assisted Gravity Drainage and other extraction technologies.

Chuck Masters of the USGS estimates that, "Taken together, these resource occurrences, in the Western Hemisphere, are approximately equal to the Identified Reserves of conventional crude oil accred to the Middle East." Authorities familiar with the resources believe that the world's ultimate reserves of unconventional oil are several times as large as those of conventional oil and will be highly profitable for companies as a result of higher prices in the 21st century. In October 2009, the USGS updated the Orinoco tar sands (Venezuela) recoverable "mean value" to 513 billion barrels (8.16×10^{10} m^3), with a 90% chance of being within the range of 380-652 billion barrels (103.7×10^9 m^3), making this area "one of the world's largest recoverable oil accumulations".

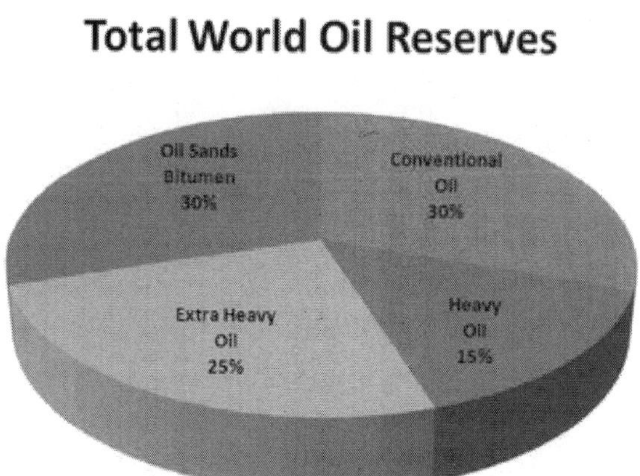

Figure : *Unconventional resources are much larger than conventional ones.*

Despite the large quantities of oil available in non-conventional sources, Matthew Simmons argues that limitations on production prevent them from becoming an effective substitute for conventional crude oil. Simmons states that "these are high energy intensity projects that can never reach high volumes" to offset significant losses from other sources. Another study claims that even under highly optimistic assumptions, “Canada’s oil sands will not prevent peak oil,” although production could reach 5,000,000 bbl/d (790,000 m^3/d) by 2030 in a “crash program” development effort.

Moreover, oil extracted from these sources typically contains contaminants such as sulphur and heavy metals that are energy-intensive to extract and can leave tailings - ponds containing hydrocarbon sludge - in some cases. The same applies to much of the Middle East’s undeveloped conventional oil reserves, much of which is heavy, viscous, and contaminated with sulphur and metals to the point of being unusable. However, recent high oil prices make these sources more financially appealing. A study by Wood Mackenzie suggests that within 15 years all the world’s extra oil supply will likely come from unconventional sources.

Synthetic Sources

Currently, two companies SASOL and Shell, have synthetic oil technology proven to work on a commercial scale. Sasol’s primary business is based on CTL (coal-to-liquid) and GTL (natural gas-to-liquid) technology, producing US$4.40 billion in revenues (FY2009). Shell has used these processes to recycle waste flare gas (usually burnt off at the refinery) into useable synthetic oil.

The US Navy is researching a modified version of the Fischer-Tropsch process that uses hydrogen and carbonates in seawater to create jet fuel on board their aircraft carriers, powered by ocean water temperature differentials. A 2003 article in *Discover* magazine claimed that thermal depolymerisation could be used to manufacture oil indefinitely, out of garbage, sewage, and agricultural waste. The article claimed that the cost of the process was $15 per barrel. A follow-up article in 2006 stated that the cost was actually $80 per barrel, because the feedstock that had previously been considered as hazardous waste now had market value.

A 2007 news bulletin published by Los Alamos Laboratory proposed that hydrogen (possibly produced using hot fluid from nuclear reactors to split water into hydrogen and oxygen) in combination with sequestered CO_2 could be used to produce methanol, which could then be converted into gasoline. The press release stated that in order for such a process to be economically feasible, gasoline prices would need to be above $4.60 “at the pump” in U.S. markets. Capital and operational costs were uncertain mostly because the costs associated with sequestering CO_2 are unknown. Another problem is that an energy source will be required for both carbon capture and water splitting processes.

Production

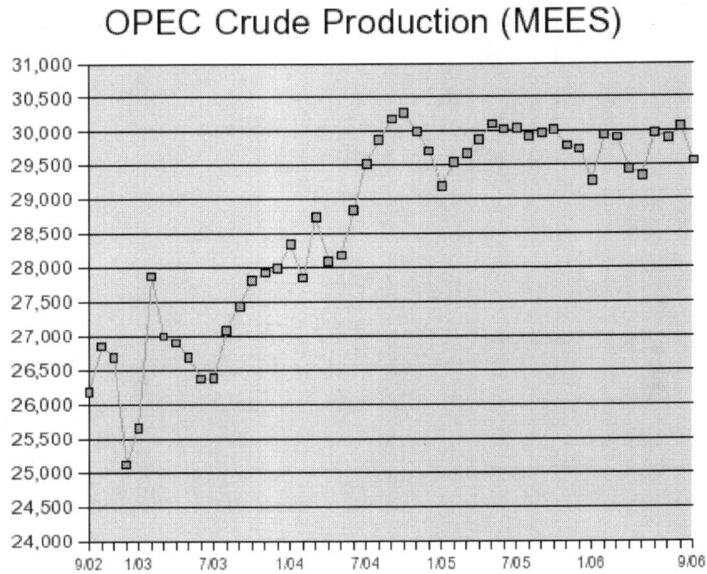

Figure : *OPEC Crude Oil Production 2002-2006 (in 1,000s barrels/day).*
Source: Middle East Economic Survey

The point in time when peak global oil production occurs defines peak oil. This is because production capacity is the main limitation of supply. Therefore, when production decreases, it becomes the main bottleneck to the petroleum supply/demand equation.

World wide oil discoveries have been less than annual production since 1980. According to several sources, worldwide production is past or near its maximum. World population has grown faster than oil production. Because of this, oil production *per capita* peaked in 1979 (preceded by a plateau during the period of 1973-1979).

The increasing investment in harder-to-reach oil is a sign of oil companies' belief in the end of easy oil. Additionally, while it is widely believed that increased oil prices spur an increase in production, an increasing number of oil industry insiders are now coming to believe that even with higher prices, oil production is unlikely to increase significantly beyond its current level. Recent work points to the difficulty of increasing production even with vastly increased investment in exploration and production, at least in mature petroleum regions. A 2008 *Journal of Energy Security* analysis of the energy return on drilling effort in the United States points to an extremely limited potential to increase production of both gas and (especially) oil. By looking at the historical response of production to variation in drilling effort, this analysis showed very little increase of production attributable to increased drilling. This was due to a tight quantitative relationship of diminishing returns with increasing drilling effort: as drilling effort increased, the energy obtained per active drill rig was reduced according to a severely diminishing power law. This fact means that even an enormous increase of drilling effort is unlikely to lead to significantly increased oil and gas production in a mature petroleum region like the United States.

Worldwide Production Trends

World oil production growth trends were flat from 2005 to 2008. According to a January 2007 International Energy Agency report, global supply (which includes biofuels, non-crude sources of petroleum, and use of strategic oil reserves, in addition to crude production) averaged 85.24 million barrels per day (13.552×10^6 m^3/d) in 2006, up 0.76 million barrels per day (121×10^3 m^3/d) (0.9%), from 2005. Average yearly gains in global supply from 1987 to 2005 were 1.2 million barrels per day (190×10^3 m^3/d) (1.7%). In 2008, the IEA drastically increased its prediction of production decline from 3.7% a year to 6.7% a year, based largely on better accounting methods, including actual research of individual oil field production throughout the world.

Oil Field Decline

Of the largest 21 fields, at least 9 are in decline. In April 2006, a Saudi Aramco spokesman admitted that its mature fields are now declining at a rate of 8% per year (with a national composite decline of about 2%). This information has been used to argue that Ghawar, which is the largest oil field in the world and responsible for approximately half of Saudi Arabia's oil production over the last 50 years, has peaked. The world's second largest oil field, the Burgan field in Kuwait, entered decline in November 2005.

According to a study of the largest 811 oilfields conducted in early 2008 by Cambridge Energy Research Associates (CERA), the average rate of field decline is 4.5% per year. The IEA stated in November 2008 that an analysis of 800 oilfields showed the decline in oil production to be 6.7% a year, and that this would grow to 8.6% in 2030. There are also projects expected to begin production within the next decade that are hoped to offset these declines. The CERA report projects a 2017 production level of over 100 million barrels per day (16×10^6 m^3/d).

Kjell Aleklett of the Association for the Study of Peak Oil and Gas agrees with their decline rates, but considers the rate of new fields coming online—100% of all projects in development, but with 30% of them experiencing delays, plus a mix of new small fields and field expansions—overly optimistic. A more rapid annual rate of decline of 5.1% in 800 of the world's largest oil fields was reported by the International Energy Agency in their World Energy Outlook 2008.

Mexico announced that its giant Cantarell Field entered depletion in March 2006, due to past overproduction. In 2000, PEMEX built the largest nitrogen plant in the world in an attempt to maintain production through nitrogen injection into the formation, but by 2006, Cantarell was declining at a rate of 13% per year.

OPEC had vowed in 2000 to maintain a production level sufficient to keep oil prices between $22–28 per barrel, but did not prove possible. In its 2007 annual report, OPEC projected that it could maintain a production level that would stabilise the price of oil at around $50–60 per barrel until 2030.

On 18 November 2007, with oil above $98 a barrel, King Abdullah of Saudi Arabia, a long-time advocate of stabilised oil prices, announced that his country would not increase production to lower prices. Saudi Arabia's inability, as the world's largest supplier, to stabilise prices through increased production during that period suggests that no nation or organisation had the spare production capacity to lower oil prices. The implication is that those major suppliers who had not yet peaked were operating at or near full capacity.

Commentators have pointed to the Jack 2 deep water test well in the Gulf of Mexico, announced 5 September 2006, as evidence that there is no imminent peak in global oil production. According to one estimate, the field could account for up to 11% of U.S. production within seven years. However, even though oil discoveries are expected after the peak oil of production is reached, the new reserves of oil will be harder to find and extract. The Jack 2 field, for instance, is more than 20,000 feet (6,100 m) under the sea floor in 7,000 feet (2,100 m) of water, requiring 8.5 kilometres (5.3 mi) of pipe to reach. Additionally, even the maximum estimate of 15 billion barrels (2.4×10^9 m^3) represents slightly less than 2 years of U.S. consumption at present levels.

Control Over Supply

Entities such as governments or cartels can reduce supply to the world market by limiting access to the supply through nationalising oil, cutting back on production, limiting drilling rights, imposing taxes, etc. International sanctions, corruption, and military conflicts can also reduce supply.

Nationalisation of Oil Supplies

Another factor affecting global oil supply is the nationalisation of oil reserves by producing nations. The nationalisation of oil occurs as countries begin to deprivatize oil production and withhold exports. Kate Dourian, Platts' Middle East or, points out that while estimates of oil reserves may vary, politics have now entered the equation of oil supply. "Some countries are becoming off limits. Major oil companies operating in Venezuela find themselves in a difficult position because of the growing nationalisation of that resource. These countries are now reluctant to share their reserves."

According to consulting firm PFC Energy, only 7% of the world's estimated oil and gas reserves are in countries that allow companies like ExxonMobil free rein. Fully 65% are in the hands of state-owned companies such as Saudi Aramco, with the rest in countries such as Russia and Venezuela, where access by Western European and North American companies is difficult. The PFC study implies political factors are limiting capacity increases in Mexico, Venezuela, Iran, Iraq, Kuwait, and Russia. Saudi Arabia is also limiting capacity expansion, but because of a self-imposed cap, unlike the other countries. As a result of not having access to countries amenable to oil exploration, ExxonMobil is not making nearly the investment in finding new oil that it did in 1981.

Cartel Influence on Supply

OPEC is an alliance between 12 diverse oil producing countries (Algeria, Angola, Ecuador, Iran, Iraq, Kuwait, Libya, Nigeria, Qatar, Saudi Arabia, the United Arab Emirates, and Venezuela) to control the supply of oil. OPEC's power was consolidated as various countries nationalised their oil holdings, and wrested decision-making away from the "Seven Sisters," (Anglo-Iranian, Socony-Vacuum, Royal Dutch Shell, Gulf, Esso, Texaco, and Socal) and created their own oil companies to control the oil. OPEC tries to influence prices by restricting production. It does this by allocating each member country a quota for production. All 12 members agree to keep prices high by producing at lower levels than they otherwise would. There is no way to verify adherence to the quota, so every member faces the same incentive to 'cheat' the cartel. Washington kept the oil flowing and gained favourable OPEC policies mainly by arming, and propping up Saudi regimes. According to some, the purpose for the second Iraq war is to break the back of OPEC and return control of the oil fields to western oil companies.

Alternately, commodities trader Raymond Learsy, author of *Over a Barrel: Breaking the Middle East Oil Cartel*, contends that OPEC has trained consumers to believe that oil is a much more finite resource than it is. To back his argument, he points to past false alarms and apparent collaboration. He also believes that peak oil analysts are conspiring with OPEC and the oil companies to create a "fabricated drama of peak oil" to drive up oil prices and profits. It is worth noting oil had risen to a little over $30/barrel at that time. A counter-argument was given in the Huffington Post after he and Steve Andrews, co-founder of ASPO, debated on CNBC in June 2007.

Increasingly Assertive Energy Producers

Think-tanks such as the Word Pensions Council (WPC) have argued that, unlike previous recessionary cycles, the price of gas could remain at a high level as Gulf Arab, Latin American and Asian governments are less inclined to accommodate the US on the supply front, which could hamper a fragile recovery in the oil-dependent nations of Europe and North America

Timing of Peak Oil

In Feb 2010 the US Joint Forces Command issued the Joint Operating Environment 2010 warning US military commands "By 2012, surplus oil production capacity could entirely disappear, and as early as 2015, the shortfall in output could reach nearly 10 million barrels per day."

"A severe energy crunch is inevitable without a massive expansion of production and refining capacity. While it is difficult to predict precisely what economic, political, and strategic effects such a shortfall might produce, it surely would reduce the prospects for growth in both the developing and developed worlds. Such an economic slowdown would exacerbate other unresolved tensions, push fragile and failing states further down the path toward collapse, and perhaps have serious economic impact on both China and India. At best, it would lead to periods of harsh economic adjustment. To what extent conservation measures, investments in alternative energy production, and efforts to expand petroleum production from tar sands and shale would mitigate such a period of adjustment is difficult to predict. One should not forget that the Great Depression spawned a number of totalitarian regimes that sought economic prosperity for their nations by ruthless conquest."

> *"Energy production and distribution infrastructure must see significant new investment if energy demand is to be satisfied at a cost compatible with economic growth and prosperity."*

> *"The discovery rate for new petroleum and gas fields over the past two decades (with the possible exception of Brazil) provides little reason for optimism that future efforts will find major new fields."*

The military warning of timing issue is graphically noted in the International Energy Agency's (IEA) World Energy Outlook 2010. The IEA graph notes that depleting conventional oil will be replaced by "fields yet to be found" and "fields yet to be developed."

M. King Hubbert initially predicted in 1974 that peak oil would occur in 1995 "if current trends continue." However, in the late 1970s and early 1980s, global oil consumption actually dropped (due to the shift to energy-efficient cars, the shift to electricity and natural gas for heating, and other factors), then rebounded to a lower level of growth in the mid 1980s. Thus oil production did not peak in 1995, and has climbed to more than double the rate initially projected. This underscores the fact that the only reliable way to identify the timing of peak oil will be in retrospect. However, predictions have been refined through the years as up-to-date information becomes more readily available, such as new reserve growth data. Predictions of the timing of peak oil include the possibilities that it has recently occurred, that it will occur shortly, or that a plateau of oil production will sustain supply for up to 100 years. None of these predictions dispute the peaking of oil production, but disagree only on when it will occur.

According to Matthew Simmons, former Chairman of Simmons & Company International and author of *Twilight in the Desert: The Coming Saudi Oil Shock and the World Economy*, "...peaking is one of these fuzzy events that you only know clearly when you see it through a rear view mirror, and by then an alternate resolution is generally too late."

Possible Consequences of Peak Oil

The wide use of fossil fuels has been one of the most important stimuli of economic growth and prosperity since the industrial revolution, allowing humans to participate in takedown, or the consumption of energy at a greater rate than it is being replaced. Some believe that when oil production decreases, human culture, and modern technological society will be forced to change drastically. The impact of peak oil will depend heavily on the rate of decline and the development and adoption of effective alternatives. If alternatives are not forthcoming, the products produced with oil (including fertilizers, detergents, solvents, adhesives, and most plastics) would become scarce and expensive.

In 2005, the United States Department of Energy published a report titled *Peaking of World Oil Production: Impacts, Mitigation, & Risk Management*. Known as the Hirsch report, it stated, "The peaking of world oil production presents the U.S. and the world with an unprecedented risk management problem. As peaking is approached, liquid fuel prices and price volatility will increase dramatically, and, without timely mitigation, the economic, social, and political costs will be unprecedented. Viable mitigation options exist on both the supply and demand sides, but to have substantial impact, they must be initiated more than a decade in advance of peaking."

High Oil Prices

Historical Oil Prices

The oil price historically was comparatively low until the 1973 oil crisis and the 1979 energy crisis when it increased more than tenfold during that six year timeframe. Even though the oil price dropped significantly in the following years, it has never come back to the previous levels.

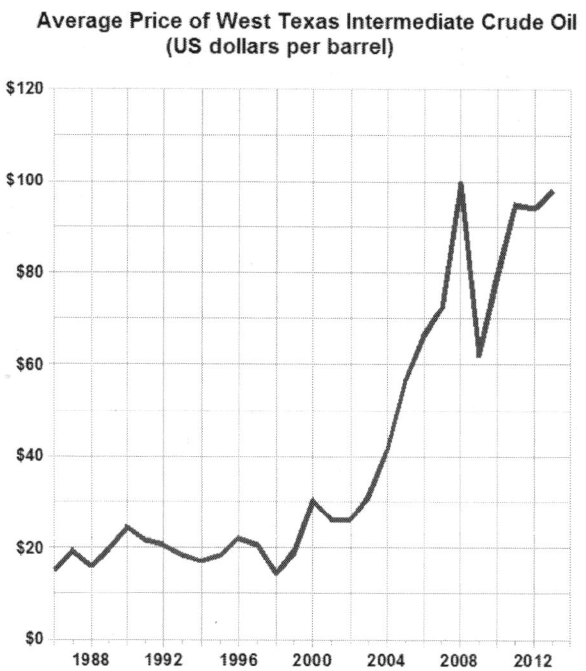

Figure : *Average Price of West Texas Intermediate Crude Oil, 1995-2011*

After almost fifteen years of relative stability, the oil price began to increase again during the 2000s until it hit historical heights of

$143 per barrel (2007 inflation adjusted dollars) on 30 June 2008. As these prices were well above those that caused the 1973 and 1979 energy crises, they have contributed to fears of an economic recession similar to that of the early 1980s. These fears were not without a basis, since the high oil prices began having an effect on the economies, as, for example, indicated by gasoline consumption drop of 0.5% in the first two months of 2008 in the United States compared to a drop of.4% total in 2007.

It is agreed that the main reason for the price spike in 2005-2008 was strong demand pressure. For example, global consumption of oil rose from 30 billion barrels (4.8×10^9 m^3) in 2004 to 31 billion in 2005. The consumption rates were far above new discoveries in the period, which had fallen to only eight billion barrels of new oil reserves in new accumulations in 2004.

Oil price increases were partially fuelled by reports that petroleum production is at or near full capacity. In June 2005, OPEC admitted that they would 'struggle' to pump enough oil to meet pricing pressures for the fourth quarter of that year. The decline in the U.S. dollar against other significant currencies from 2007 to 2008 is also cited as a significant reason for the oil price increases, as the dollar lost approximately 14% of its value against the Euro from May 2007 to May 2008.

Besides supply and demand pressures, at times security related factors may have contributed to increases in prices, including the War on Terror, missile launches in North Korea, the Crisis between Israel and Lebanon, nuclear brinkmanship between the U.S. and Iran, and reports from the U.S. Department of Energy and others showing a decline in petroleum reserves.

Effects of Rising Oil Prices

In the past, the price of oil has led to economic recessions, such as the 1973 and 1979 energy crises. The effect the price of oil has on an economy is known as a price shock. In many European countries, which have high taxes on fuels, such price shocks could potentially be mitigated somewhat by temporarily or permanently suspending the taxes as fuel costs rise. This method of softening price shocks is less useful in countries with much lower gas taxes, such as the United States.

Researchers at the Stanford Energy Modelling Forum found that the economy can adjust to steady, gradual increases in the price of crude better than wild lurches.

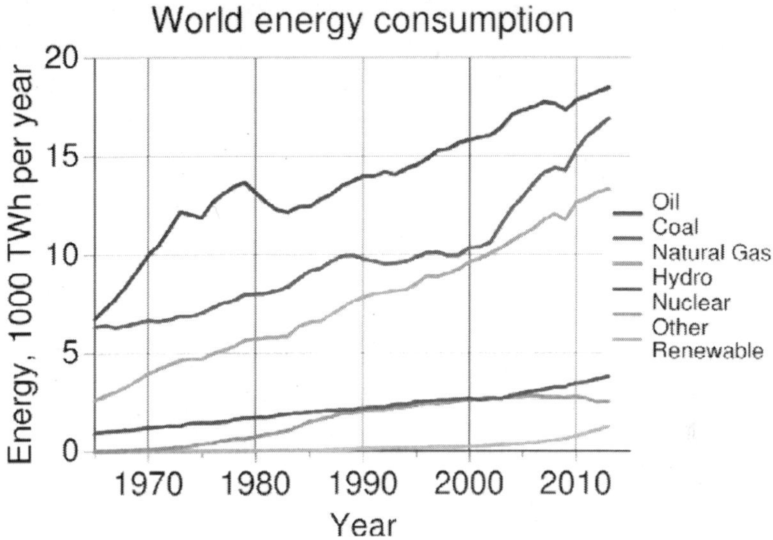

Figure: *World consumption of primary energy by energy type.*

Some economists predict that a substitution effect will spur demand for alternate energy sources, such as coal or liquefied natural gas. This substitution can only be temporary, as coal and natural gas are finite resources as well.

Prior to the run-up in fuel prices, many motorists opted for larger, less fuel-efficient sport utility vehicles and full-sized pickups in the United States, Canada, and other countries. This trend has been reversing due to sustained high prices of fuel. The September 2005 sales data for all vehicle vendors indicated SUV sales dropped while small cars sales increased. Hybrid and diesel vehicles are also gaining in popularity.

In 2008, a report by Cambridge Energy Research Associates stated that 2007 had been the year of peak gasoline usage in the United States, and that record energy prices would cause an "enduring shift" in energy consumption practices. According to the report, in April gas consumption had been lower than a year before for the sixth straight month, suggesting 2008 would be the first year U.S. gasoline usage declined in 17 years. The total miles driven in the U.S. peaked in 2006.

The Export Land Model states that after peak oil petroleum exporting countries will be forced to reduce their exports more quickly than their production decreases because of internal demand growth. Countries that rely on imported petroleum will therefore be affected earlier and more dramatically than exporting countries. Mexico is already in this situation. Internal consumption grew by 5.9% in 2006 in the five biggest exporting countries, and their exports declined by over 3%. It was estimated that by 2010 internal demand would decrease worldwide exports by 2,500,000 barrels per day (400,000 m^3/d).

Canadian economist Jeff Rubin has stated that high oil prices will likely result in increased consumption in developed countries through partial manufacturing de-globalisation of trade. Manufacturing production would move closer to the end consumer to minimise transportation network costs, and therefore a demand decoupling from Gross Domestic Product would occur. Higher oil prices would lead to increased freighting costs and consequently, the manufacturing industry would move back to the developed countries since freight costs would outweigh the current economic wage advantage of developing countries. Chinese Export data released on 10 March 2012 confirmed a deep slowdown in exports, as China entered an unexpectedly large trade deficit.

Economic research carried out by the International Monetary Fund puts overall price elasticity of demand for oil at -0.025 short term and -0.093 long term.

Long-term Effects on Lifestyle

A majority of Americans live in suburbs, a type of low-density settlement designed around universal personal automobile use. Commentators such as James Howard Kunstler argue that because over 90% of transportation in the U.S. relies on oil, the suburbs' reliance on the automobile is an unsustainable living arrangement. Peak oil would leave many Americans unable to afford petroleum based fuel for their cars, and force them to use bicycles or electric vehicles. Additional options include telecommuting, moving to rural areas, or moving to higher density areas, where walking and public transportation are more viable options. In the latter two cases, suburbia may become the "slums of the future." The issues of petroleum supply and demand is also a concern for growing cities in developing countries (where urban areas are expected to absorb most of the world's projected 2.3 billion population increase by 2050). Stressing the energy component of future development plans is seen as an important goal.

Methods that have been suggested for mitigating these urban and suburban issues include the use of non-petroleum vehicles such as electric cars, battery electric vehicles, transit-oriented development, Car-free Cities, bicycles, new trains, new pedestrianism, smart growth, shared space, urban consolidation, urban villages, and New Urbanism.

An extensive 2009 report by the United States National Research Council of the Academy of Sciences, commissioned by the United States

Congress, stated six main findings. First, that compact development is likely to reduce "Vehicle Miles Travelled" (VMT) throughout the country. Second, that doubling residential density in a given area could reduce VMT by as much as 25% if coupled with measures such as increased employment density and improved public transportation. Third, that higher density, mixed-use developments would produce both direct reductions in CO_2 emissions (from less driving), and indirect reductions (such as from lower amounts of materials used per housing unit, higher efficiency climate control, longer vehicle lifespans, and higher efficiency delivery of goods and services).

Fourth, that although short term reductions in energy use and CO_2 emissions would be modest, that these reductions would grow over time. Fifth, that a major obstacle to more compact development in the United States is political resistance from local zoning regulators, which would hamper efforts by state and regional governments to participate in land-use planning. Sixth, the committee agreed that changes in development that would alter driving patterns and building efficiency would have various secondary costs and benefits that are difficult to quantify. The report recommends that policies supporting compact development (and especially its ability to reduce driving, energy use, and CO_2 emissions) should be encouraged.

An economic theory that has been proposed as a remedy is the introduction of a steady state economy. Such a system would include a tax shifting from income to depleting natural resources (and pollution), as well as the limitation of advertising that stimulates demand and population growth. It also includes the institution of policies that move away from globalisation and toward localisation to conserve energy resources, provide local jobs, and maintain local decision-making authority. Zoning policies would be adjusted to promote resource conservation and eliminate sprawl.

Mitigation

To avoid the serious social and economic implications a global decline in oil production could entail, the 2005 Hirsch report emphasized the need to find alternatives, at least ten to twenty years before the peak, and to phase out the use of petroleum over that time. This was similar to a plan proposed for Sweden that same year. Such mitigation could include energy conservation, fuel substitution, and the use of unconventional oil. Because mitigation can reduce the use of traditional petroleum sources, it can also affect the timing of peak oil and the shape of the Hubbert curve.

Positive Aspects of Peak Oil

Some observers opine that peak oil should be viewed as a positive event. Many such critics reason that if the price of oil rises high enough, the use of alternative clean fuels could help control pollution from fossil fuel use, and mitigate global warming. Permaculture, particularly as expressed in the work of Australian David Holmgren, and others, sees peak oil as holding tremendous potential for positive change, assuming countries act with foresight. The rebuilding of local food networks, energy production, and the general implementation of 'energy descent culture' are argued to be ethical responses to the acknowledgment of finite fossil resources.

The Transition Towns movement, started in Totnes, Devon and spread internationally by "The Transition Handbook" (Rob Hopkins), sees the restructuring of society for more local resilience and ecological stewardship as a natural response to the combination of peak oil and climate change.

Criticisms

Some do not agree with peak oil, at least as it has been presented by Matthew Simmons. The president of Royal Dutch Shell's U.S. operations John Hofmeister, while agreeing that conventional oil production will soon start to decline, has criticized Simmons's analysis for being "overly focused on a single country: Saudi Arabia, the world's largest exporter and OPEC swing producer."

He also points to the large reserves at the U.S. outer continental shelf, which holds an estimated 100 billion barrels (16×10^9 m^3) of oil and natural gas. As things stand, however, only 15% of those reserves are currently exploitable, a good part of that off the coasts of Louisiana, Alabama, Mississippi, and Texas. Hofmeister also contends that Simmons erred in excluding unconventional sources of oil such as the oil sands of Canada, where Shell is already active. The Canadian oil sands—a natural combination of sand, water, and oil found largely in Alberta and Saskatchewan—is believed to contain one trillion barrels of oil. Another trillion barrels are also said to be trapped in rocks in Colorado, Utah, and Wyoming, but are in the form of oil shale. These particular reserves present major environmental, social, and economic obstacles to recovery. Hofmeister also claims that if oil companies were allowed to drill more in the United States enough to produce another 2 million barrels per day (320×10^3 m^3/d), oil and gas prices would not be as high as they are in the later part of the 2000 to 2010 decade. He thinks that high energy prices are causing social unrest similar to levels surrounding the Rodney King riots.

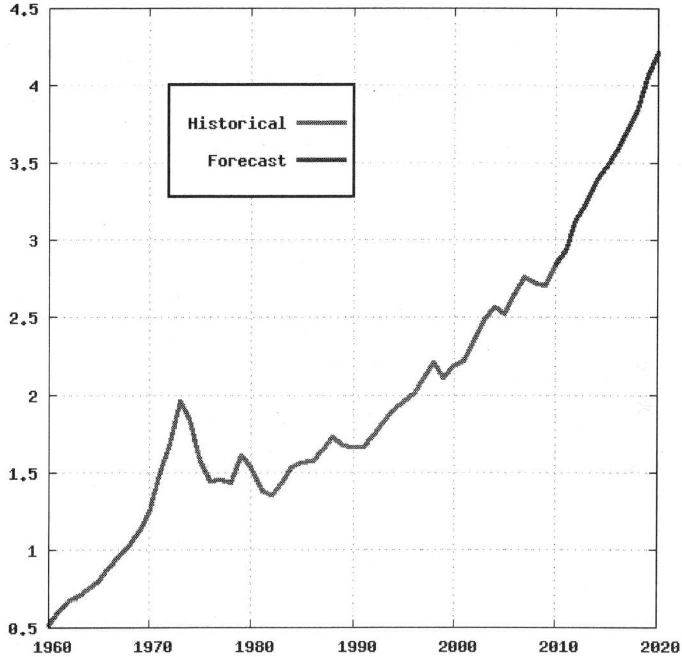

Figure: *Canadian conventional oil production peaked in 1973, but non-conventional oil sands production continues to increase total oil production*

Dr. Christoph Rühl, chief economist of BP, repeatedly uttered strong doubts about the peak oil hypothesis:

Physical peak oil, which I have no reason to accept as a valid statement either on theoretical, scientific or ideological grounds, would be insensitive to prices. (...)In fact the whole hypothesis of peak oil – which is that there is a certain amount of oil in the ground, consumed at a certain rate, and then it's finished – does not react to anything.... (Global Warming) is likely to be more of a natural limit than all these peak oil theories combined. (...) Peak oil has been predicted for 150 years. It has never happened, and it will stay this way.

According to Rühl, the main limitations for oil availability are "above ground" and are to be found in the availability of staff, expertise, technology, investment security, money and last but not least in global warming. The oil question is about price and not the basic availability. His views are shared by Daniel Yergin of CERA, who added that the recent high price phase might add to a future demise of the oil industry - not of lack of resources or an apocalyptic shock but the timely and smooth setup of alternatives.

Clive Mather, CEO of Shell Canada, said the Earth's supply of hydrocarbons is almost infinite, referring to hydrocarbons in oil sands. Engineer Peter Huber believes the Canadian oil sands can fuel all of humanity's needs for over 100 years.

Industry blogger Steve Maley echoed some of the points of Yergin, Rühl, Mather and Hofmeister.

Oil and Gas Law in the United States

Oil and gas law in the United States is the branch of law that pertains to the acquisition and ownership rights in oil and gas both under the soil before discovery and after its capture, and adjudication regarding those rights.

Overview

The law regulating oil and gas ownership in the U.S. generally differs significantly from laws in Europe because oil and gas are often owned privately in the U.S. as opposed to being owned by the national government in many other countries.

Jurisdiction

In the U.S., extraction of oil and gas is generally regulated by the individual states through statutes and common law. Federal and constitutional law apply as well.

Ownership

In the United States, oil and gas rights to a particular parcel may be owned by private individuals, corporations, Indian tribes, or by local, state, or federal governments. Oil and gas rights extend vertically downward from the property line. Unless explicitly separated by a deed, oil and gas rights are owned by the surface landowner. Once severed from surface ownership, oil and gas rights may be bought, sold, or transferred, like other real estate property. Oil and gas rights offshore are owned by either the state or federal government and leased to oil companies for development. The tidelands controversy involve the limits of state ownership. Although oil and gas laws vary by state, the laws regarding ownership prior to, at, and after extraction are nearly universal. An owner of real estate also owns the minerals underneath the surface, unless the minerals are severed under a previous deed or an agreement.

Prior to and at Extraction

Unless mineral rights are severed, whoever owns the fee of the soil owns everything below the surface, limited by the extent of the surface rights (*Del Monte Mining & Milling Co. v. Last Chance Mining & Milling Co.*). Because oil and gas are fluids, they may flow in the subsurface across property boundaries. In this way, an operator may permissibly extract oil and gas from beneath the land of another, if the extraction is lawfully conducted on his own property (*Kelly v. Ohio Oil Co.*). An operator may not, however, angle a well to penetrate beneath property not owned by or leased to him. The two conflicting legal doctrines covering oil and gas extraction are the rule of capture, and the correlative rights doctrine. Which of the doctrines applies in a particular case depends on state law, which varies considerably from state to state, or in the case of the federal offshore zone, on US federal law. The rule of capture gives land owners an incentive to pump out oil as quickly as possible to capture the oil of their neighbours. Such practice may deplete the gas pressure needed to force oil out of the ground. Government agencies such as the Texas Railroad Commission therefore regulate extraction by individual owners.

Ownership of Extracted Oil and Gas

Refined hydrocarbons that escape into the ground are not subject to the law of capture, unless evidence shows that the refiner abandoned them. *Champlin Exploration, Inc. v. Western Bridge & Steel Col, Inc.* Extracted oil and gas which are subsequently stored in underground reservoirs are considered as personal property, rather than as an interest in real estate. *Texas American Energy Corporation v. Citizens Fidelity Bank & Trust Company.*

Lease

Oil and gas producing companies do not always own the land they drill on. Most often the company (the *lessee*) leases the mineral rights from the owner (the *lessor*). Major points in a lease include the description of the property, the term (duration), and the payments to the lessor.

Lessees of mineral rights have a right of reasonable access to leased land to explore, develop, and transport minerals (*Hunt Oil Co. v. Kerbaugh*), unless the lease specifies otherwise (a "no-surface access" lease).

Term of the Lease

A lease remains in effect for a certain period of time, called the primary term, as long as the lessee pays the annual rental. The lease expires after the primary term, unless drilling or oil and gas production has started on the lease. If production is established, the lease will remain in effect past the primary term, as long as the lease continuously produces oil or gas. The lease can however, be revived by virtue of delay rentals. Delay rentals are fees paid to the lessor, to delay production or commencement of drilling, without terminating the lease. There are other clauses that also, revive the lease. (Mentioned below).

To commence drilling a well under the habendum clause means that substantial preparations for such drilling has to be undertaken, as long as such measures have been commenced in good faith and with due diligence (*Breaux v. Apache Oil Corp.*). The habendum clause sets out these terms, as well as most significantly, identifying the parties to the transaction and their interests in the conveyed real property.

An oil and gas lease generally includes a *force majeure* clause. Such agreement relieves the lessee from liability for breach, if the party's performance is impeded as the result of a natural cause that could not have been anticipated or prevented. This Act of God must completely prevent performance and must be unanticipated. Courts often construe this clause very strictly and rarely enforce it. For example, a tornado preventing performance in Oklahoma would not trigger the *force majeure* clause, since tornadoes are a common occurrence in Oklahoma. The Responsible Federal Oil and Gas Lease Act (2008), also called the "Use It or Lose It" bill (HR 6251 IH), proposed prohibiting the Secretary of the Interior from issuing new federal oil and gas leases to holders of existing leases who do not either diligently develop the lands subject to such existing leases or relinquish such leases. This bill failed to pass in the House of Representatives.

Payments

Payments to the lessor typically take three forms: bonus, rental, and royalty. The bonus is an up-front payment made at the time the lease takes effect. The rental is an annual payment, usually made until such time as the property begins producing oil or gas in commercial quantities. The royalty is a portion of the value of any oil or gas produced from the lease.

In an "unless-delay rental" lease, a lessee agrees to pay delay rentals so long as the lessee is not drilling on the property. An "unless" oil and gas lease terminates automatically, if the lessee fails to drill within the specified time or pay the delay rentals as called for in the lease (*Schwartemnerger v. Hunt Trust Estate*)

Contract

Oil and gas contracts have nuances which differ from standard contracts. For example, when an assignment of an oil and gas lease expressly provides that any extension or renewal of the lease is subject to an overriding royalty, a new lease that is substantially similar to the first lease and procured by the assignee during the term of the first lease, is regarded, as a matter of law, as an extension of renewal of the first lease. *Reynolds-Rexwinkle Oil, Inc. v. Petex, Inc.*

Statutes can override agreements made by parties. For instance, a statute may void an agreement to indemnify a construction worker as to liability for death or bodily injury incurred on an oil well, regardless of the indemnitee's negligence, without affecting the validity of an insurance contract. It affirms the right of an individual party to obtain insurance, not to protect the interests of the indemnitee. Amoco Production Co. v. Action Well Service, Inc. *These suits for negligence are typically brought by drilling site workers known as roustabouts.*

In Education and Practice

Law school classes teaching oil and gas law generally require that students first take a class in property and contract law. In Texas and Wyoming, oil and gas law is tested on the bar exam.

Oil and gas law practitioners usually fall into three broad categories. First, oil and gas companies usually have in-house attorneys that advise the company of its rights and the legal issues. These attorneys are usually assisted by landmen, who examine property titles land oil and gas rights and acquire property for the company. Landmen may be lawyers themselves. Second, practitioners may represent private parties. When an oil company attempts to obtain land from a private party, a party may retain counsel to be better informed of his or her rights and to negotiate a favourable bargain with the oil company. Last, oil and gas attorneys work for federal and state governments that oversee energy and environmental policy and land acquisitions.

Abiogenic Petroleum Origin

Abiogenic petroleum origin is a hypothesis that was proposed as an alternative to theory of biological petroleum origin. It was relatively popular in the past, but it became largely forgotten at the end of the 20th century after it failed to predict the location of new wells.

The abiogenic hypothesis argues that petroleum was formed from deep carbon deposits, perhaps dating to the formation of the Earth. Supporters of the abiogenic hypothesis suggest that a great deal more petroleum exists on Earth than commonly thought, and that petroleum may originate from carbon-bearing fluids that migrate upward from the mantle. The presence of methane on Saturn's moon Titan and in the atmospheres of Jupiter, Saturn, Uranus and Neptune is cited as evidence of the formation of hydrocarbons without biology.

The biogenic theory for petroleum was first proposed by Georg Agricola in the 16th century and various abiogenic hypotheses were proposed in the 19th century, most notably by Alexander von Humboldt, the Russian chemist Dmitri Mendeleev and the French chemist Marcellin Berthelot. Abiogenic hypotheses were revived in the last half of the 20th century by Russian and Ukrainian scientists, who had little influence outside the Soviet Union because most of their research was published in their native languages. The theory was re-defined and made popular in the West by Thomas Gold, who published all his research in English.

Although the abiogenic hypothesis was accepted by many geologists in the former Soviet Union, it allegedly fell out of favour because it never made any useful prediction for the discovery of oil deposits. Most geologists now consider the abiogenic formation of petroleum scientifically unsupported. The abiogenic origin of petroleum has also recently been reviewed in detail by Glasby, who raises a number of objections, including that there is no direct evidence to date of abiogenic *petroleum* (liquid crude oil and long-chain hydrocarbon compounds).

It has been recently discovered that thermophilic bacteria, in the sea bottom and in cooling magma, produce methane and hydrocarbon gases, but studies indicate they are not produced in commercially significant quantities (i.e. in extracted hydrocarbon gases, the median abiogenic hydrocarbon content is 0.02%, or 1 part in 5,000).

History of Abiogenic Hypothesis

The abiogenic theory for the origin of petroleum is usually traced to the early part of the 19th century. The hypothesis was developed well before the field of organic chemistry, much less that of biochemistry, was established so the chemical nature of the petroleum was not known. Absent intellectual framework of organic and biological chemistry, abiologic theories were inevitable. In the early 19th century, Phlogiston theory was the dominant model for explaining chemical phenomena. Furthermore, the formal study of paleontology had only started in the early 19th century. It is within this scientifically primitive but changing environment that theories on the origin of petroleum originated.

Alexander von Humboldt was the first to propose an inorganic abiogenic hypothesis for petroleum formation after he observed petroleum springs in the Bay of Cumaux (Cumaná) on the northeast coast of Venezuela. In 1804 he is quoted as saying, "petroleum is the product of a distillation from great depth and issues from the primitive rocks beneath which the forces of all volcanic action lie." Abraham

Gottlob Werner and the proponents of neptunism in the 18th century believed basaltic sills to be solidified oils or bitumen. While these notions have been proven unfounded, the basic idea that petroleum is associated with magmatism persisted. Other prominent proponents of what would become the abiogenic hypothesis included Mendeleev and Berthelot.

Russian geologist Nikolai Alexandrovitch Kudryavtsev proposed the modern abiotic hypothesis of petroleum in 1951. On the basis of his analysis of the Athabasca Oil Sands in Alberta, Canada, he concluded that no "source rocks" could form the enormous volume of hydrocarbons, and that therefore the most plausible explanation is abiotic deep petroleum. However, humic coals have since been proposed for the source rocks. Kudryavtsev's work was continued by Petr N. Kropotkin, Vladimir B. Porfir'ev, Emmanuil B. Chekaliuk, Vladilen A. Krayushkin, Georgi E. Boyko, Georgi I. Voitov, Grygori N. Dolenko, Iona V. Greenberg, Nikolai S. Beskrovny, and Victor F. Linetsky.

Astronomer Thomas Gold was the most prominent proponent of the abiogenic hypothesis in the West until his death in 2004. Currently, Jack Kenney of Gas Resources Corporation is a prominent proponent in the West.

State of Current Research

Little research is directed on establishing abiogenic petroleum or methane, although the Carnegie Institution for Science have found that ethane and heavier hydrocarbons can be synthesized under conditions of the upper mantle. Research mostly related to astrobiology and the deep microbial biosphere and serpentinite reactions, however, continue to provide insight into the contribution of abiogenic hydrocarbons into petroleum accumulations.

- rock porosity and migration pathways for abiogenic petroleum
- ocean floor hydrothermal vents as in the Lost City hydrothermal field;
- Mud volcanoes and the volatile contents of deep pelagic oozes and deep formation brines
- mantle peridotite serpentinization reactions and other natural Fischer-Tropsch analogs
- Primordial hydrocarbons in meteorites, comets, asteroids and the solid bodies of the solar system
- Primordial or ancient sources of hydrocarbons or carbon in Earth

- Primordial hydrocarbons formed from hydrolysis of metal carbides of the iron peak of cosmic elemental abundance (Cr, Fe, Ni, V, Mn, Co)
- isotopic studies of groundwater reservoirs, sedimentary cements, formation gases and the composition of the noble gases and nitrogen in many oil fields
- the geochemistry of petroleum and the presence of trace metals related to Earth's mantle (Ni, V, Cd, As, Pb, Zn, Hg and others)

Similarly, research into the deep microbial hypothesis of hydrocarbon generation is advancing as part of the attempt to investigate the concept of panspermia and astrobiology, specifically using deep microbial life as an analog for life on Mars. Research applicable to deep microbial petroleum theories includes

- Research into how to sample deep reservoirs and rocks without contamination
- Sampling deep rocks and measuring chemistry and biological activity
- Possible energy sources and metabolic pathways which may be used in a deep biosphere
- Investigations into the reworking primordial hydrocarbons by bacteria and their effects on carbon isotope fractionation

A 2006 review article by Glasby presented arguments against the abiogenic origin of petroleum on a number of counts.

Foundations of the Hypotheses

Within the mantle, carbon may exist as hydrocarbons, chiefly methane and as elemental carbon, carbon dioxide and carbonates.

The abiotic hypothesis is that the full suite of hydrocarbons found in petroleum can be generated in the mantle by abiogenic processes, and these hydrocarbons can migrate out of the mantle into the crust until they escape to the surface or are trapped by impermeable strata, forming petroleum reservoirs.

Abiogenic theories reject the supposition that certain molecules found within petroleum, known as biomarkers, are indicative of the biological origin of petroleum.

They contend that these molecules mostly come from microbes feeding on petroleum in its upward migration through the crust, that some of them are found in meteorites, which have presumably never contacted living material, and that some can be generated abiogenically by plausible reactions in petroleum.

The hypothesis is founded primarily upon:

Proponents	***Item***
Gold	The presence of methane on other planets, meteors, moons and comets Gold, Kenney Proposed mechanisms of abiotically chemically synthesizing hydrocarbons within the mantle Kudryavtsev,
Gold	Hydrocarbon-rich areas tend to be hydrocarbon-rich at many different levels Kudryavtsev, Gold Petroleum and methane deposits are found in large patterns related to deep-seated large-scale structural features of the crust rather than to the patchwork of sedimentary deposits
Gold	Interpretations of the chemical and isotopic composition of natural petroleum Kudryavtsev,
Gold	The presence of oil and methane within non-sedimentary rocks upon the Earth
Gold	The existence of methane hydrate deposits
Gold	Perceived ambiguity in some assumptions and key evidence used in the orthodox biogenic petroleum theories
Gold	Bituminous coal creation is based upon deep hydrocarbon seeps
Gold	Surface carbon budget and oxygen levels stable over geologic time scales
Kudryavtsev, Gold	Biogenic theories do not explain some hydrocarbon deposit characteristics
Szatmari	The distribution of metals in crude oils fits better with upper serpentinized mantle, primitive mantle and chondrite patterns than oceanic and continental crust, and show no correlation with sea water
Gold	The association of hydrocarbons with helium, a noble gas

Conventional Theories

According to generally accepted theory, petroleum is derived from ancient biomass. The theory was initially based on the isolation of molecules from petroleum that closely resemble known biomolecules.

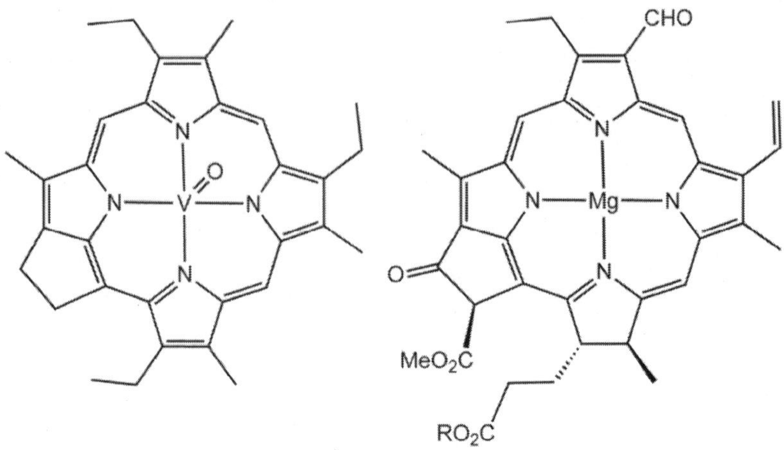

Figure : *Structure of a biomarker extracted from petroleum and simplified structure of chlorophyll a.*

Most petroleum geologists prefer theories of oil formation, which holds that oil originated in shallow seas as vast quantities of marine plankton or plant materials died and sank into the mud. Under the resulting anaerobic conditions organic compounds remained in a reduced state where anaerobic bacteria converted the lipids (fats, oils and waxes) into a waxy substance called kerogen.

As the source rock was buried deeper, overburden pressure raised temperatures into the *oil window*, between 80 and 180 °C. Most of the organic compounds degraded into the straight-chain hydrocarbons that comprise most of petroleum. This process is called the generation *kitchen.* Once crude oil formed, it became very fluid and migrated upward through the rock strata. This process is called oil expulsion. Eventually it was either trapped in an oil reservoir or oil escaped to the surface and was biodegraded by soil bacteria.

Oil buried deeper entered the "gas window" of more than 160 °C and was converted into natural gas by thermal cracking. Thus, theory predicts that no oil will be found below a certain depth, only unassociated gas. At greater depths, even natural gas would be pyrolysed.

Proposed Mechanisms of Abiogenic Petroleum

Primordial Deposits

Thomas Gold's work was focused on hydrocarbon deposits of primordial origin. Meteorites are believed to represent the major composition of material from which the Earth was formed. Some meteorites, such as carbonaceous chondrites, contain carbonaceous material. If a large amount of this material is still within the Earth, it could have been leaking upward for billions of years. The thermodynamic conditions within the mantle would allow many hydrocarbon molecules to be at equilibrium under high pressure and high temperature. Although molecules in these conditions may disassociate, resulting fragments would be reformed due to the pressure. An average equilibrium of various molecules would exist depending upon conditions and the carbon-hydrogen ratio of the material.

Creation within the Mantle

Russian researchers concluded that hydrocarbon mixes would be created within the mantle. Experiments under high temperatures and pressures produced many hydrocarbons, including n-alkanes through $C_{10}H_{22}$, from iron oxide, calcium carbonate, and water. Because such materials are in the mantle and in subducted crust, there is no requirement that all hydrocarbons be produced from primordial deposits.

Hydrogen Generation

Hydrogen gas and water have been found more than 6 kilometres deep in the upper crust, including in the Siljan Ring boreholes and the Kola Superdeep Borehole. Data from the western United States suggests that aquifers from near the surface may extend to depths of 10 to 20 km. Hydrogen gas can be created by water reacting with silicates, quartz and feldspar, in temperatures in the 25 to 270 °C range. These minerals are common in crustal rocks such as granite. Hydrogen may react with dissolved carbon compounds in water to form methane and higher carbon compounds.

One reaction not involving silicates which can create hydrogen is:

Ferrous oxide + Water → Magnetite + hydrogen

$$3FeO + H_2O \rightarrow Fe_3O_4 + H_2$$

The above reaction operates best at low pressures. At pressures greater than 5 GPa almost no hydrogen is created.

Serpentinite Mechanism

One proposed mechanism by which abiogenic petroleum is formed was first proposed by the Ukrainian scientist, Prof. Emmanuil B. Chekaliuk in 1967. He proposed that petroleum could be formed at high temperatures and pressures from inorganic carbon in the form of carbon dioxide, hydrogen and/or methane.

This mechanism is supported by several lines of evidence which are accepted by modern scientific literature. This involves synthesis of oil within the crust via catalysis by chemically reductive rocks. A proposed mechanism for the formation of inorganic hydrocarbons is via natural analogs of the Fischer-Tropsch process known as the *serpentinite mechanism* or the serpentinite process.

$$CH_4 + \frac{1}{2}O_2 \rightarrow 2H_2 + CO$$

$$(2n+1)H_2 + nCO \rightarrow C_nH_{2n+2} + nH_2O$$

Serpentinites are ideal rocks to host this process as they are formed from peridotites and dunites, rocks which contain greater than 80% olivine and usually a percentage of Fe-Ti spinel minerals. Most olivines also contain high nickel concentrations (up to several percent) and may also contain chromite or chromium as a contaminant in olivine, providing the needed transition metals.

However, serpentinite synthesis and spinel cracking reactions require hydrothermal alteration of pristine peridotite-dunite, which is a finite process intrinsically related to metamorphism, and further, requires significant addition of water. Serpentinite is unstable at mantle temperatures and is readily dehydrated to granulite, amphibolite, talc–schist and even eclogite. This suggests that methanogenesis in the presence of serpentinites is restricted in space and time to mid-ocean ridges and upper levels of subduction zones. However, water has been found as deep as 12 km, so water-based reactions are dependent upon the local conditions. Oil being created by this process in intracratonic regions is limited by the materials and temperature.

Serpentinite Synthesis

A chemical basis for the abiotic petroleum process is the serpentinization of peridotite, beginning with methanogenesis via hydrolysis of olivine into serpentine in the presence of carbon dioxide. Olivine, composed of Forsterite and Fayalite metamorphoses into serpentine, magnetite and silica by the following reactions, with silica from fayalite decomposition (reaction 1a) feeding into the forsterite reaction (1b).

Reaction 1a:

Fayalite + water → Magnetite + aqueous silica + Hydrogen

$$3Fe_2SiO_4 + 2H_2O \rightarrow 2Fe_3O_4 + 3SiO_2 + 2H_2$$

Reaction 1b:

Forsterite + aqueous silica → Serpentinite

$$3Mg_2SiO_4 + SiO_2 + 4H_2O \rightarrow 2Mg_3Si_2O_5(OH)_4$$

When this reaction occurs in the presence of dissolved carbon dioxide (carbonic acid) at temperatures above 500 °C Reaction 2a takes place.

Reaction 2a:

Olivine + Water + Carbonic acid → Serpentine + Magnetite + Methane

$$(Fe,Mg)_2SiO_4 + nH_2O + CO_2 \rightarrow Mg_3Si_2O_5(OH)_4 + Fe_3O_4 + CH_4$$

or, in balanced form: $18Mg_2SiO_4 + 6Fe_2SiO_4 + 26H_2O + CO_2 \rightarrow$

$$12Mg_3Si_2O_5(OH)_4 + 4Fe_3O_4 + CH_4$$

However, reaction 2(b) is just as likely, and supported by the presence of abundant talc-carbonate schists and magnesite stringer veins in many serpentinised peridotites;

Reaction 2b:

Olivine + Water + Carbonic acid → Serpentine + Magnetite + Magnesite + Silica

$$(Fe,Mg)_2SiO_4 + nH_2O + CO_2 \rightarrow Mg_3Si_2O_5(OH)_4 + Fe_3O_4 + MgCO_3 + SiO_2$$

The upgrading of methane to higher n-alkane hydrocarbons is via dehydrogenation of methane in the presence of catalyst transition metals (e.g. Fe, Ni). This can be termed spinel hydrolysis.

Spinel Polymerisation Mechanism

Magnetite, chromite and ilmenite are Fe-spinel group minerals found in many rocks but rarely as a major component in non-ultramafic rocks. In these rocks, high concentrations of magmatic magnetite, chromite and ilmenite provide a reduced matrix which may allow abiotic cracking of methane to higher hydrocarbons during hydrothermal events. Chemically reduced rocks are required to drive this reaction and high temperatures are required to allow methane to be polymerised to ethane. Note that reaction 1a, above, also creates magnetite.

Reaction 3:

Methane + Magnetite → Ethane + Hematite

$$nCH_4 + nFe_3O_4 + nH_2O \rightarrow C_2H_6 + Fe_2O_3 + HCO_3 + H^+$$

Reaction 3 results in n-alkane hydrocarbons, including linear saturated hydrocarbons, alcohols, aldehydes, ketones, aromatics, and cyclic compounds.

Carbonate Decomposition

Calcium carbonate may decompose at around 500 °C through the following reaction:

Reaction 5:

Hydrogen + Calcium carbonate → Methane + Calcium oxide + Water

$$4H_2 + CaCO_3 \rightarrow CH_4 + CaO + 2H_2O$$

Note that CaO (lime) is not a mineral species found within natural rocks. Whilst this reaction is possible, it is not plausible.

Evidence of Abiogenic Mechanisms

- Calculations by J.F. Kenney using scaled particle theory for a simplified perturbed hard-chain, (a statistical mechanical model) predict that methane compressed to 30 or 40 kbar at 1000 °C (conditions in the mantle) is relatively unstable in relation to higher hydrocarbons. However, these calculations do not include methane pyrolisis yielding amorphous carbon and hydrogen, which is recognised as the prevalent reaction at high temperatures.
- Experiments in diamond anvil high pressure cells have resulted in partial conversion of methane and inorganic carbonates into light hydrocarbons.

Biotic (Microbial) Hydrocarbons

The "deep biotic petroleum hypothesis", similar to the abiogenic petroleum origin hypothesis, holds that not all petroleum deposits within the Earth's rocks can be explained purely according to the orthodox view of petroleum geology. Thomas Gold used the term *the deep hot biosphere* to describe the microbes which live underground.

This hypothesis is different from biogenic oil in that the role of deep-dwelling microbes is a biological source for oil which is not of a sedimentary origin and is not sourced from surface carbon. Deep microbial life is only a contaminant of primordial hydrocarbons. Parts of microbes yield molecules as biomarkers.

Deep biotic oil is considered to be formed as a byproduct of the life cycle of deep microbes. Shallow biotic oil is considered to be formed as a byproduct of the life cycles of shallow microbes.

Microbial Biomarkers

Thomas Gold, in a 1999 book, cited the discovery of thermophile bacteria in the Earth's crust as new support for the postulate that these bacteria could explain the existence of certain biomarkers in extracted petroleum. A rebuttal of biogenic origins based on biomarkers has been offered by Kenney, et al. (2001).

Isotopic Evidence: Methane is ubiquitous in crustal fluid and gas. Research continues to attempt to characterise crustal sources of methane as biogenic or abiogenic using carbon isotope fractionation of observed gases (Lollar & Sherwood 2006). There are few clear examples of abiogenic methane-ethane-butane, as the same processes favour enrichment of light isotopes in all chemical reactions, whether organic or inorganic. δC of methane overlaps that of inorganic carbonate and graphite in the crust, which are heavily depleted in C, and attain this by isotopic fractionation during metamorphic reactions.

One argument for abiogenic oil cites the high carbon depletion of methane as stemming from the observed carbon isotope depletion with depth in the crust. However, diamonds, which are definitively of mantle origin, are not as depleted as methane, which implies that methane carbon isotope fractionation is not controlled by mantle values.

Commercially extractable concentrations of helium (greater than 0.3%) are present in natural gas from the Panhandle-Hugoton fields in the USA, as well as from some Algerian and Russian gas fields.

Helium trapped within most petroleum occurrences, such as the occurrence in Texas, is of a distinctly crustal character with an *Ra* ratio of less than 0.0001 that of the atmosphere.

The Chimaera gas seep, near Antalya (SW Turkey), new and thorough molecular and isotopic analyses including methane (~87% v/v; D13C1 from -7.9 to -12.3 ‰; D13D1 from -119 to -124 ‰), light alkanes (C2+C3+C4+C5 = 0.5%; C6+: 0.07%; D13C2 from -24.2 to -26.5 ‰; D13C3 from -25.5 to -27 ‰), hydrogen (7.5 to 11 %), carbon dioxide (0.01-0.07%; D13CCO2: -15 ‰), helium (~80 ppmv; R/Ra: 0.41) and nitrogen (2-4.9%; D15N from -2 to -2.8 ‰) converge to indicate that the seep releases a mixture of organic thermogenic gas, related to mature Type III kerogen occurring in Paleozoic and Mesozoic organic rich sedimentary rocks, and abiogenic gas produced by low temperature serpentinization in the Tekirova ophiolitic unit.

Biomarker Chemicals

Certain chemicals found in naturally occurring petroleum contain chemical and structural similarities to compounds found within many living organisms. These include terpenoids, terpenes, pristane, phytane, cholestane, chlorins and porphyrins, which are large, chelating molecules in the same family as heme and chlorophyll. Materials which suggest certain biological processes include tetracyclic diterpane and oleanane.

The presence of these chemicals in crude oil is assumed to be as a result of the inclusion of biological material in the oil. This is predicated upon the theory that these chemicals are released by kerogen during the production of hydrocarbon oils, as these are chemicals highly resistant to degradation and plausible chemical paths have been studied. Abiotic defenders state that biomarkers get into oil during its way up as it gets in touch with ancient fossils. However a more plausible explanation is that biomarkers are traces of biological molecules from bacteria (archaea) that feed on primordial hydrocarbons and die in that environment. For example, hopanoids are just parts of the bacterial cell wall present in oil as contaminant.

Trace Metals

Nickel (Ni), vanadium (V), lead (Pb), arsenic (As), cadmium (Cd), mercury (Hg) and others metals frequently occur in oils. Some heavy crude oils, such as Venezuelan heavy crude have up to 45% vanadium pentoxide content in their ash, high enough that it is a commercial source for vanadium. Abiotic supporters argue that these metals are common in Earth's mantle, but relatively high contents of nickel, vanadium, lead and arsenic can be usually found in almost all marine sediments.

Analysis of 22 trace elements in oils correlate significantly better with chondrite, serpentinized fertile mantle peridotite, and the primitive mantle than with oceanic or continental crust, and shows no correlation with seawater.

Reduced Carbon

Sir Robert Robinson studied the chemical makeup of natural petroleum oils in great detail, and concluded that they were mostly far too hydrogen-rich to be a likely product of the decay of plant debris, assuming a dual origin for Earth hydrocarbons. However, several processes which generate hydrogen could supply kerogen hydrogenation which is compatible with conventional petroleum generation theories.

Olefins, the unsaturated hydrocarbons, would have been expected to predominate by far in any material that was derived in that way. He also wrote: "Petroleum... seems to be a primordial hydrocarbon mixture into which bio-products have been added."

This has however been demonstrated later to be a misunderstanding by Robinson, related to the fact that only short duration experiments were available to him. Olefins are thermally very unstable (that is why natural petroleum normally does not contain such compounds) and in laboratory experiments that last more than a few hours, the olefins are no longer present.

The presence of low-oxygen and hydroxyl-poor hydrocarbons in natural living media is supported by the presence of natural waxes (n=30+), oils (n=20+) and lipids in both plant matter and animal matter, for instance fats in phytoplankton, zooplankton and so on. These oils and waxes, however, occur in quantities too small to significantly affect the overall hydrogen/carbon ratio of biological materials. However, after the discovery of highly aliphatic biopolymers in algae, and that oil generating kerogen essentially represent concentrates of such materials, no theoretical problem exists anymore. Also, the millions of source rock samples that have been analysed for petroleum yield by the petroleum industry have confirmed the large quantities of petroleum found in sedimentary basins.

Field Observations

Proponents sometimes cite occurrences of abiotic petroleum in comercial amounts, in the oil wells in offshore Vietnam, Eugene Island block 330 oil field, and the Dnieper-Donets Basin, but its origin can be explained by biotic theories. Modern geogologists think that commercially-profitable deposits of abiotic petroleum *could* be found, it's just that all current deposits can already be explained by biotic theories, and no current deposit has convincing evidence that it originated from abiotic sources.

The Russian-Ukrainian school saw evidence of their theory in the fact that some oil reservoirs exist in non-sedimentary rocks such as granite, metamorphic or porous volcanic rocks. However, critics note that non-sedimentary rocks served as reservoirs for biologically-originated oil expelled from nearby sedimentary source rock through common migration or re-migration mechanisms.

The following observations have been commonly been used to argue for the abiogenic hypothesis, however all the observations of actual petroleum can also be fully explained by conventional petroleum formation theories.

Lost City Hydrothermal Vent Field

The Lost City Hydrothermal Vent Field was determined to have abiogenic hydrocarbon production. Proskurowski et al. wrote, "Radiocarbon evidence rules out seawater bicarbonate as the carbon source for FTT reactions, suggesting that a mantle-derived inorganic carbon source is leached from the host rocks. Our findings illustrate that the abiotic synthesis of hydrocarbons in nature may occur in the presence of ultramafic rocks, water, and moderate amounts of heat."

Siljan Ring, Sweden

The Siljan Ring meteorite crater, Sweden, was proposed by Thomas Gold as the most likely place to test the hypothesis because it was one of the few places in the world where the granite basement was cracked sufficiently (by meteorite impact) to allow oil to seep up from the mantle; furthermore it is infilled with a relatively thin veneer of sediment, which was sufficient to trap any abiogenic oil, but was modelled as not having been subjected to the heat and pressure conditions (known as the "oil window") normally required to create biogenic oil. However, some geochemists concluded by geochemical analysis that the oil in the seeps came from the organic-rich Ordivician Tretaspis shale, where it was heated by the meteorite impact.

In 1986-1990 The Gravberg-1 borehole was drilled through the deepest rock in the Siljan Ring in which proponents had hoped to find hydrocarbon reservoirs. It stopped at the depth of 6,800 m due to drilling problems, after private investors spent $40 millions. Some eighty barrels of magnetite paste and hydrocarbon-bearing sludge were recovered from the well; Gold maintained that the hydrocarbons were chemically different from, and not derived from, those added to the borehole, but analyses showed that the hydrocarbons were derived from the diesel fuel-based drilling fluid used in the drilling. This well also sampled over 13,000 feet (4,000 m) of methane-bearing inclusions. In 1991-1992, a second borehole, Stenberg-1, was drilled a few miles away to a depth of 6,500 m, finding similar results. Again, no abiotic hydrocarbons were found.

Bacterial Mats

Direct observation of bacterial mats and fracture-fill carbonate and humin of bacterial origin in deep boreholes in Iran, Australia, Sweden and Canada are also taken as evidence for the abiogenic origin of petroleum.

Example Proposed Abiogenic Methane Deposits

Panhandle-Hugoton field (Anadarko Basin) in the south-central United States is the most important gas field with commercial helium content.

The B¡ch Hϕ oil field in Vietnam has been proposed as an example of abiogenic oil because it is 4,000 m of fractured basement granite, at a depth of 5,000 m. However, others argue that it contains biogenic oil which leaked into the basement horst from conventional source rocks within the Cuu Long basin.

A major component of mantle-derived carbon is indicated in commercial gas reservoirs in the Pannonian and Vienna basins of Hungary and Austria.

Natural gas pools interpreted as being mantle-derived are the Shengli Field and Songliao Basin, northeastern China.

The Chimaera gas seep, near Çýralý, Antalya (southwest Turkey), has been continuously active for millennia and it is known to be the source of the first Olympic fire in the Hellenistic period. On the basis of chemical composition and isotopic analysis, the Chimaera gas is said to be about half biogenic and half abiogenic gas, the largest emission of biogenic methane discovered; deep and pressurised gas accumulations necessary to sustain the gas flow for millennia, posited to be from an inorganic source, may be present. Local geology of Chimaera flames, at exact position of flames, reveals contact between serpentinized ophiolite and carbonate rocks. Fischer-Tropsch process can be suitable reaction to form hydrocarbons gases.

The Geological Argument for Abiogenic Oil

Given the known occurrence of methane and the probable catalysis of methane into higher atomic weight hydrocarbon molecules, the abiogenic hypothesis considers the following to be key observations in support;

- The serpentinite synthesis, graphite synthesis and spinel catalysation models prove the process is viable
- The likelihood that abiogenic oil seeping up from the mantle is trapped beneath sediments which effectively seal mantle-tapping faults
- Mass-balance calculations for supergiant oilfields which argue that the calculated source rock could not have supplied the reservoir with the known accumulation of oil, implying deep recharge (Kudryavtsev, 1951)
- The presence of hydrocarbons encapsulated in diamonds

Incidental Evidence

The proponents of abiogenic oil use several arguments which draw on a variety of natural phenomena in order to support the hypothesis

- The modelling of some researchers which shows the Earth was accreted at relatively low temperature, thereby perhaps preserving primordial carbon deposits within the mantle, to drive abiogenic hydrocarbon production

- The presence of methane within the gases and fluids of mid-ocean ridge spreading centre hydrothermal fields

The Geological Argument Against

Key arguments against chemical reactions, such as the serpentinite mechanism, as being the major source of hydrocarbon deposits within the crust are;

- The lack of available pore space within rocks as depth increases
- This is contradicted by numerous studies which have documented the existence of hydrologic systems operating over a range of scales and at all depths in the continental crust.
- The lack of any hydrocarbon within the crystalline shield areas of the major cratons, especially around key deep seated structures which are predicted to host oil by the abiogenic hypothesis.
- Limited evidence that major serpentinite belts underlie continental sedimentary basins which host oil
- Lack of conclusive proof that carbon isotope fractionation observed in crustal methane sources is entirely of abiogenic origin (Lollar et al. 2006)
- Mass balance problems of supplying enough carbon dioxide to serpentinite within the metamorphic event before the peridotite is fully reacted to serpentinite
- Drilling of the Siljan Ring failed to find commercial quantities of oil, thus providing a counter example to Kudryavtsev's Rule and failing to locate the predicted abiogenic oil.
- Helium in the Siljan Gravberg-1 well was depleted in He and not consistent with a mantle origin
- The Gravberg-1 well only produced 84 barrels (13.4 m^3) of oil, which later was shown to derive from organic additives, lubricants and mud used in the drilling process.
- The distribution of sedimentary basins is caused by plate tectonics, with sedimentary basins forming on either side of a volcanic arc, which explains the distribution of oil within these sedimentary basins
- Kudryavtsev's Rule has been explained for oil and gas (not coal): Gas deposits which are below oil deposits can be created from

that oil or its source rocks. Because natural gas is less dense than oil, as kerogen and hydrocarbons are generating gas the gas fills the top of the available space. Oil is forced down, and can reach the spill point where oil leaks around the edge(s) of the formation and flows upward. If the original formation becomes completely filled with gas then all the oil will have leaked above the original location.

- Ubiquitous presence of diamondoids in natural hydrocarbons such as oil, gas and condensates are composed of carbon from biological sources, unlike the carbon found in normal diamonds.

Arguments Against the Incidental Evidence

- Gas ruptures during earthquakes are more likely to be sourced from biogenic methane generated in unconsolidated sediment from existing organic matter, released by earthquake liquefaction of the reservoir during tremors
- The presence of methane hydrate is arguably produced by bacterial action upon organic detritus falling from the littoral zone and trapped in the depth due to pressure and temperature
- The likelihood of vast concentrations of methane in the mantle is very slim, given mantle xenoliths have negligible methane in their fluid inclusions; conventional plate tectonics explains deep focus quakes better, and the extreme confining pressures invalidate the hypothesis of gas pockets causing quakes
- Further evidence is the presence of diamond within kimberlites and lamproites which sample the mantle depths proposed as being the source region of mantle methane (by Gold et al.).

Chapter 3

Integrated Operations

In the Petroleum industry, Integrated operations (IO) refers to new work processes and ways of performing oil and gas exploration and production, which has been facilitated by new information and communication technology. Multi-discipline collaboration in plant operation is one example. IO has in a sense also taken the form of a movement for renewal of the oil and gas industry. In short IO is collaboration with production in focus.

Contents of the Term

The most striking part of IO has been the use of always-on videoconference rooms between offshore platforms and land-based offices. This includes broadband connections for sharing of data and video-surveillance of the platform. This has made it possible to move some personnel onshore and use the existing human resources more efficiently. Instead of having e.g. an expert in geology on duty at every platform, the expert may be stationed on land and be available for consultation for several offshore platforms. It's also possible for a team at an office in a different time zone to be consulting the night-shift of the platform, so that no land-based workers need work at night.

Splitting the team between land and sea demands new work processes, which together with ICT is the two main focus points for IO. Tools like videoconferencing and 3D-visualisation also creates an opportunity for new, more cross-discipline cooperations. For instance, a shared 3D-visualisation may be tailored to each member of the group, so that the geologist gets a visualisation of the geological structures while the drilling engineer focuses on visualising the well. Here, real-time measurements from the well are important but the downhole bandwidth has previously been very restricted. Improvements in bandwidth, better measurement devices, better aggregation and visualisation of this information and improved models that simulate the rock formations and wellbore currently all feed on each other. An important task where all these improvements play together is real-time production optimisation.

In the process industry in general, the term is used to describe the increased cooperation, independent of location, between operators, maintenance personnel, electricians, production management as well as business management and suppliers to provide a more streamlined plant operation.

By deploying IO, the petroleum industry draws on lessons from the process industry. This can be seen in a larger focus on the whole production chain and management ideas imported from the production and process industry. A prominent idea in this regard is real-time optimisation of the whole value chain, from long term management of the oil reservoir, through capacity allocations in pipe networks and calculations of the net present value of the produced oil.

A focus on the whole production chain is also seen in debates about how to organise people in an IO organisation, with frequent calls for breaking down the Information silos in the oil companies. A large oil company is typically organised in functional silos corresponding to disciplines such as drilling, production and reservoir management. This is regarded as inefficient by the IO movement, pointing out that the activities in any well or field by any of the silos will involve or affect all of the others. While some companies focus on their inhouse management structure, others also emphasize the integration and coordination of outside suppliers and collaborators in offshore-operations. For instance, it is pointed out that the oil and gas industry is lagging behind other industries in terms of Operational intelligence.

Ideas and theories that IO management and work processes build on will be familiar from operations research, knowledge management and continual improvement as well as information systems and business transformation. This is perhaps most evident in the repeated referral to "people, process and technology" in IO discussions. As bullet-points this mirror many of the aforementioned fields.

Incentives

Common to most companies is that IO leads to cost savings as fewer people are stationed offshore and an increased efficiency. Lower costs, more efficient reservoir management and fewer mistakes during well drilling will in turn raise profits and make more oil fields economically viable. IO comes at a time when the oil industry is faced with more "brown fields", also referred to as "tail production", where the cost of extracting the oil will be higher than its market value, unless major improvements in technology and work processes are made. It has been estimated that deployment of IO could produce 300 billion NOK of added value to the Norwegian continental shelf alone. On a longer time-scale, working onshore control and monitoring of the oil production may become a necessity as new fields at deeper waters are based purely on unmanned sub-sea facilities.

Moving jobs onshore has also been touted as a way to keep and make better use of an aging workforce, which is regarded as a challenge by western oil and gas companies. As the average age of the industry workforce is increasing with many nearing retirement, IO is being leveraged for knowledge sharing and training of younger workforce. More comfortable onshore jobs together with "high-tech" tools has also been fronted as a way to recruit young workers into an industry that is seen as "unsexy", "lowtech" and difficult to combine with a normal family life.

Critique

The security aspect of reducing the offshore workforce has been raised. Will on-site experience be lost and can familiarity with the platform and its processes be attained from an onshore office? The new working environment in any case demands changes to HSE routines. Some of the challenges also include clear role and responsibility definitions and clarifications between the onshore & offshore personnel. Who in a given situation has the authority to take decisions, the onsite or the offshore staff. The increased integration of the offshore facilities with the onshore office environment and outside collaborators also expose work-critical ICT-infrastructure to the internet and the hazards of everyday ICT. As for the efficiency aspect, some criticize the onshore-offshore collaboration for creating a more bureaucratic working environment.

Naming Conventions

Both the exact terms and the content used to describe IO vary between companies. The oil company Shell has traditionally branded the term Smart Fields, which was an extension of Smart Wells that only referred to remote-controlled well-valves. BP uses Field of the future, Chevron has i-field and Schlumberger terms it Digital Energy. The latter term, understood as referring to oil and gas, is adopted in the title of the digital energy journal. This term could have several meanings, as GE Digital Energy for instance, do not appear to use it in the IO sense.

Other terms include e-Field, i-Field, Digital Oilfield, Intelligent Oilfield, Field of the future and Intelligent Energy. Integrated operations has been the preferred term by Statoil, the Norwegian Oil Industry Association (OLF), a professional body and employer's association for oil and supplier companies and vendors such as ABB. Intelligent Energy is the dominant term in publications revolving around the biannual SPE Intelligent Energy conference, which has been one of the major conferences for the IO movement, along with the annual IO Science and Practice conference which obviously supports the IO term.

Energy Policy of the United Kingdom

The current energy policy of the United Kingdom is set out in the Energy White Paper of May 2007 and Low Carbon Transition Plan of July 2009, building on previous work including the 2003 Energy White Paper and the Energy Review Report in 2006. It is led by the Department of Energy and Climate Change, currently headed by Ed Davey.

Overview

The 2007 White Paper: "Meeting the Energy Challenge" sets out the Government's international and domestic energy strategy to address the long term energy challenges faced by the UK, and to deliver 4 key policy goals:

1. To put the UK on a path to cut carbon dioxide emissions by some 60% by about 2050, with real progress by 2020;
2. To maintain reliable energy supplies;
3. To promote competitive markets in the UK and beyond, helping to raise the rate of sustainable economic growth and to improve productivity; and
4. To ensure that every home is adequately and affordably heated.

The scope of energy policy includes the production and distribution of electricity, transport fuel usage, and means of heating (significantly Natural Gas). The policy recognises: "Energy is essential in almost every aspect of our lives and for the success of our economy. We face two long-term energy challenges:

- Tackling climate change by reducing carbon dioxide emissions both within the UK and abroad; and

- Ensuring secure, clean and affordable energy as we become increasingly dependent on imported fuel."

The policy also recognises that the UK will need around 30-35GW of new electricity generation capacity over the next two decades as many of the UK's current coal and nuclear power stations, built in the 1960s and 1970s, reach the end of their lives and are set to close.

The 2006 Energy Review reintroduced the prospect of new nuclear power stations in the UK. However, following a judicial review requested by Greenpeace, on February 15, 2007 elements of the 2006 Energy Review were ruled 'seriously flawed', and 'not merely inadequate but also misleading'. As a result, plans to build a new generation of nuclear power plants were ruled illegal at that time. In response, the Government ran "The Future of Nuclear Power" consultation from May to October 2007. The Government's response to the consultation conclusions, published in January 2008, state "set against the challenges of climate change and security of supply, the evidence in support of new nuclear power stations is compelling."

The January 2008 Energy Bill updates the legislative framework in the UK to reflect their current policy towards the energy market and the challenges faced on climate change and security of supply. Key elements of the bill address nuclear, carbon capture and storage, renewables, and offshore gas and oil. A framework to encourage investment in nuclear power within a new regulatory environment was simultaneously published in the January 2008 Nuclear White Paper.

In October 2008, the Government created the Department of Energy and Climate Change to bring together energy policy (previously with the Department for Business, Enterprise and Regulatory Reform), and climate change mitigation policy (previously with the Department for Environment, Food and Rural Affairs).

Scotland

Though energy policy is an area reserved to the UK government under the Scotland Act 1998 that established devolved government for Scotland, the Scottish Government has an energy policy for Scotland at variance with UK policy, and has planning powers to enable it to put its policy priorities into effect.

1980s Market Liberalisation

Under the Conservatives during the 1980s and 1990s, Government policy was one of market liberalisation linked to the privatisation of state controlled energy companies and the dismantling of the Department of Energy.

As a consequence, Government no longer has the ability to directly control the energy markets. Regulation is now carried out through the Office of Gas and Electricity Markets (OFGEM), while energy policy is largely limited to influencing the operation of the market.

Such influence is exerted through taxation (such as North Sea Oil Tax 5), subsidy (such as the Renewables Obligation), incentives, planning controls, the underwriting of liabilities (such as those carried by the Nuclear Decommissioning Authority), grants, and funding for research.

Energy Market

A *Research and Markets* review estimated the 2006 total market value of UK inland energy consumption was GBP 130.73bn. Consumption by the energy sector was valued at GBP 28.73bn, while the value of consumption by the non-energy sector was GBP 128.2bn, with transport being the largest component of the non-energy sector. The UK is currently proposing a wide ranging reforms of it's electricity market, including measures such as contracts for difference for generators and a capacity market to ensure security of supply in the latter half of this decade.

Primary Energy Sources

Historically a country emphasising its coal, nuclear and off-shore natural gas production, the United Kingdom is currently in transition to become a net energy importer.

In the year 2007 the percentage of primary energy derived from major sources was as follows:

- Oil: 38.0%
- Natural gas: 37.7%
- Coal: 16.7%
- Nuclear power: 5.8%
- Renewable: 1.8%.

Coal

Coal usage may be expected to decline steadily because of eroding cost advantages and pressure to reduce sulphur and carbon (carbon dioxide) emissions, notwithstanding ongoing subsidy policies designed to retain jobs in the coal mining industry. Future coal usage is highly dependent on legislative drivers on emissions and the need to have security of supply. Whilst the costs of burning coal with desulphurisation and carbon capture facilities is greatly increased, it is still being actively considered as part of the UK energy strategy due to large domestic reserves, higher price stability than natural gas and reduced capital expenditure and construction time for plant compared to nuclear power.

The 2002 Energy Review concluded that the option of new investment in clean coal technology (through carbon sequestration) needed to be kept open, and that practical measures should be taken to do this 6.

Gas

During the 1980s and early 1990s, there was a massive expansion in gas-fired generation capacity, known as the Dash for Gas. The rapidity of construction of gas-fired plants (compared to coal-fired or nuclear plants) was especially attractive due to the high interest rates of the period. Natural gas looks set to take a smaller part in providing future UK energy needs. Domestic production from the North Sea gas fields continues to lessen. And despite investment to enhance pipelines and storage of imported natural gas (Mostly from Norway) there is a reluctance to allow too great a reliance on Russia and its gas exports for energy needs.

By the year 2021, North Sea oil and natural gas production is predicted to slip 75 percent from 2005 levels to less than one million barrels per year. Oil and coal reserves for all of Europe are among the most tenuous in the developed world: for example, Europe's reserves to annual consumption ratio stands at 3.0, perilously low by world standards.

Nuclear Power

Following the UK Government's January 2008 decision to support the building of new nuclear power stations, EDF announced that it plans to open four new plants in the UK by 2017. EON and Centrica have also shown interest in building further plants. However, Scotland's First Minister Alex Salmond has said there is "no chance" of more nuclear power stations being built in Scotland as the Scottish Government is opposed.

Renewables

From the mid 1990s renewable energy began to contribute to the electricity generated in the United Kingdom, adding to a small hydroelectricity generating capacity. Renewable energy sources provided for 6.7 per cent of the electricity generated in the United Kingdom in 2009, rising to 9.6% in the second quarter of 2011.

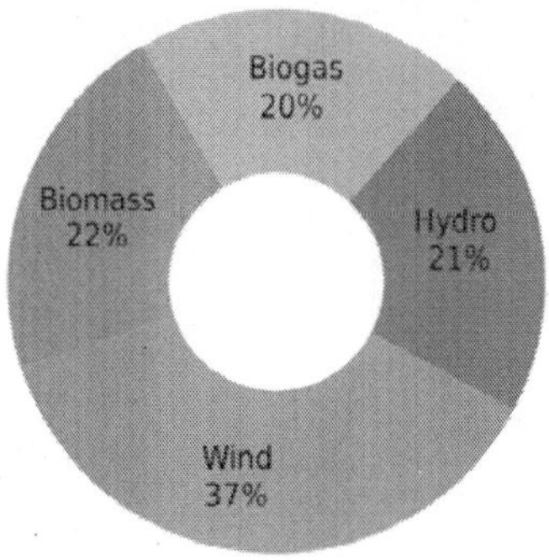

***Figure :** Electricity generation from renewable sources in the United Kingdom (2009)*

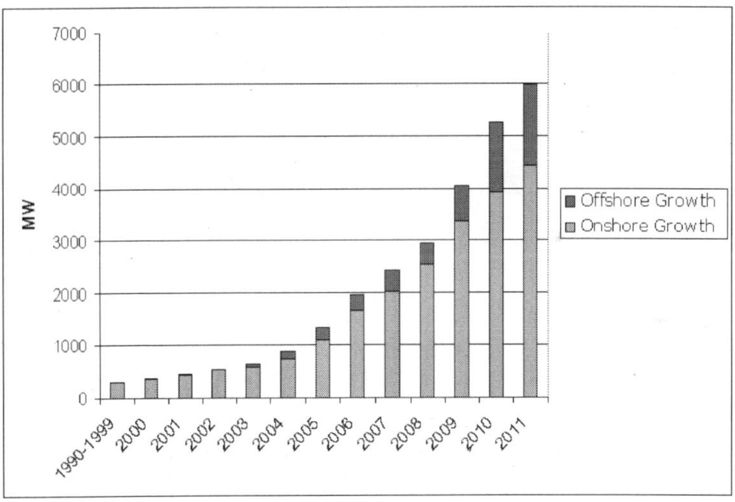

***Figure :** UK installed wind power capacity 1990–2011*

By mid-2011, the installed capacity of wind power in the United Kingdom was over 5.7 gigawatts and the UK is ranked as the world's eighth largest producer of wind power. Wind power is expected to continue growing in the UK for the foreseeable future, Renewable UK estimates that more than 2 GW of capacity will be deployed per year for the next five years. Within the UK, wind power is the second largest source of renewable energy after biomass.

The UK Government's goal for renewable energy production is to produce 20% of electricity in the UK by the year 2020. The 2002 Energy Review 7 set a target of 10% to be in place by 2010/2011. The target was increased to 15% by 2015 and most recently the 2006 Energy Review further set a target of 20% by 2020.

For Scotland, the Scottish Executive has a target of generating 17% to 18% of electricity from renewables by 2010 8, rising to 80% by 2020 9. Renewables located in Scotland count towards both the Scottish target and to the overall target for the UK.

Energy end Usage

Year 2005 UK end use energy percentage is approximately:

- Transport: 35%
- Space heating: 26%
- Industrial: 10%
- Water heating: 8%
- Lighting/small electrics: 6%

There is a steady increase of fuel usage driven by an increasingly affluent and mobile population, so that fuel use increased by ten percent in the decade ending 2000. This trend is expected to be mitigated by increased percentage of more efficient diesel and hybrid vehicles.

United Kingdom space and hot water heating consume a greater share of end use compared to the USA and more mild southern European or tropical climates. With regard to building and planning issues affecting energy use, the UK has developed guidance documents to promote energy conservation through local councils, especially as set forth in Part L of the Building Regulations (Conservation of Fuel and power). The associated document. Part 2B, addresses commercial uses, and is generally complete as to heating issues; the guidance is lacking on lighting issues, except with guidelines for local switching of lighting controls. In particular there are no standards set forth for illumination levels, and over-illumination is one of the most significant unneeded costs of commercial energy use.

Carbon Emissions Reduction

Joining over 170 other nations the UK has committed to reduction of carbon dioxide emissions, with consequent constraints to its energy policy. The UK produces four percent of the world's greenhouse gases as of 2003, compared to 23 percent by the US and 20 percent for the rest of Europe. The long term reduction goal for carbon emissions is 80 percent decrease by the year 2050. A scheme of trading for carbon emission crs has been developed in Europe that will allow some of the reduction to arise from economic transactions.

Road transport emissions reduction has been stimulated since 1999 by the banding of Vehicle Excise Duty. Bands for new vehicles are based on the results of a laboratory test, designed to calculate the theoretical potential emissions of the vehicle in grammes of CO_2 per kilometre travelled, under ideal conditions.

Average carbon emissions fell from 192 to 172 grams/mile between 1995 and 2004. Aviation fuel is not regulated under the Kyoto Protocol, so that if the UK is successful in carbon emission reduction, aviation will constitute 25 percent of UK generated greenhouse gases by the year 2030.

The UK government has one project in the planning stage for natural gas fed power generation with carbon capture by seawater. This facility is contemplated at Peterhead, Scotland, a relatively remote exposure to the North Sea.

Climate Change Act

On March 13, 2007, a draft Climate Change Bill was published following cross-party pressure over several years, led by environmental groups. The Act puts in place a framework to achieve a mandatory 80% cut in the UK's carbon emissions by 2050 (compared to 1990 levels), with an intermediate target of between 26% and 32% by 2020. The Bill was passed into law in November 2008. With its passing the United Kingdom became the first country in the world to set such a long-range and significant carbon reduction target into law, or to create such a legally binding framework.

The Committee on Climate Change, whose powers are invested by Part 2 of the Act, was formally launched in December 2008 with Lord Adair Turner as its chair.

In April 2009 the Government set a requirement for a 34% cut in emissions by 2020, in line with the recommendations of the Committee on Climate Change, and announced that details of how this would be achieved would be published in the summer.

Analysis

Prof Kevin Anderson raised concern about the growing effect of air transport on the climate in a paper and a presentation in 2008. Anderson holds a Chair in Energy and Climate Change at the School of Mechanical, Aerospace and Civil Engineering at the University of Manchester in the UK. He has pointed out that even at a reduced annual rate of increase in UK passenger air travel and with the government's targeted emissions reductions in other energy use sectors, by 2030 aviation would be causing 70% of the UK's allowable CO_2 emissions.

The UK Low Carbon Transition Plan

Published on July 15, 2009, the UK Low Carbon Transition Plan details the actions to be taken to cut carbon emissions by 34% by 2020, based on 1990 levels (of which 21% had been achieved at the time of publication). As a result, by 2020 is it envisaged that:

- Over 1.2 million people will be employed in green jobs.
- The efficiency of 7 million homes will have been upgraded, with over 1.5 million of them generating renewable energy.
- 40% of electricity will be generated from low carbon sources (renewables, nuclear power and clean coal).
- Gas imports will be 50% lower than would otherwise have been the case.
- The average new car will emit 40% less carbon compared to 2009 levels.

Renewable Energy

The first targets for renewable energy, 5% of by the end of 2003 and 10% by 2010 'subject to the cost to consumers being acceptable' were set by Helen Liddell in 2000. The 2003 Energy White Paper revised the established goals for UK renewable sources to 10% of electricity generation by 2010 and 20% by 2020, though by 2007 the then Energy Minister Malcolm Wicks indicated that by 2020 the figures has 'got to be somewhere between 10% and 15%'.

Subsequently the Low Carbon Transition Plan of 2009 made clear that by 2020 the UK would need to produce 30% of its electricity from renewables.

Although renewable energy sources have not played a major role in the UK historically, there is potential for significant use of tidal power and wind energy (both on-shore and off-shore) as recognised by formal UK policies, including the *Energy White Paper* and directives to councils in the form of PPS 22. The Renewables Obligation acts as the central mechanism for support of renewable sources of electricity in the UK, and should provide subsidies approaching one billion pounds sterling per annum by 2010. A number of other grants and smaller support mechanisms aim to support less established renewables. In addition, renewables have been exempted from the Climate Change Levy that affects all other energy sources.

The amount of renewable generation added in the year 2004 was 250 megawatts and 500 megawatts in 2005. There is also a program established for micro-generation (less than 50 KWe (kilowatt electrical) or 45 KWt (kilowatt thermal) from a low carbon source) as well as a solar voltaic program. By comparison both Germany and Japan have photovoltaic (solar cell) programmes much larger than the installed base in the UK. Hydroelectric energy is not a viable option for most of the UK due to terrain and lack of force of rivers.

The government has established a goal of five percent of the total transport fuel that must be from renewable sources (e.g. ethanol, biofuel) by the year 2010 under the Renewable Transport Fuel Obligation. This goal may be ambitious, without the necessary infrastructure and paucity of research on appropriate UK crops, but import from France might be a realistic option (based upon the French wine lake).

In 2005 British Sugar announced that it will build the UK's first ethanol biofuel production facility, using British grown sugar beet as the feed stock. The plant in Norfolk will produce 55,000 metric tonnes of ethanol annually when it is completed in the first quarter of 2007. However it has been argued that even using all the UK's set-aside land to grow biofuel crops would provide for less than seven percent of the UK's present transport fuel usage.

UK electricity generation from renewable energy in 2009 - Total is 26,999 GWh and a further 10,968 GWh produced in heat, mainly from biomass boilers (7% of the electricity generated)

Source	***GWh***	***%***
hydro	5357	19.8
biomass	4487	16.6
landfill gas	5170	19.1
municipal solid waste combustion & C&I	2389	8.8
onshore wind	7483	27.7
offshore wind	2109	7.8
wave power	1	0.004
tidal stream	3	0.01
solar photovoltaics	20	0.07

Fuel Poverty

Reducing occurrence of fuel poverty (defined as households paying over ten percent of income for heating costs) is one of the four basic goals of UK energy policy. In the prior decade substantial progress has been made on this goal, but primarily due to government subsidies to the poor rather than through fundamental change of home design or improved energy pricing.

The following national programs have been specifically instrumental in such progress: Winter Fuel Payment, Child Tax Cr and Pension Cr. Some benefits have resulted from the Warm Front Scheme in England, the Central Heating Programme in Scotland and the Home Energy Efficiency Scheme in Wales. These latter programs provide economic incentives for physical improvement in insulation, etc.

Energy White Paper

The 2007 Energy White Paper: *Meeting the Energy Challenge* was published on May 23, 2007. The 2007 White Paper outlines the Government's international and domestic strategy for responding to two main challenges:

- cutting carbon emissions to tackle global warming
- ensuring secure, clean and affordable energy as imports replace declining production from North Sea oil and gas

It seeks to do this in a way that is consistent with its four energy policy goals:

- cutting the UK's carbon dioxide emissions by some 60% by about 2050, with real progress by 2020;
- maintaining the reliability of energy supplies;
- promoting competitive markets in the UK and beyond, helping to raise the rate of sustainable economic growth and to improve productivity; and
- ensuring that every home is adequately and affordably heated.

The paper anticipates that it will be necessary to install 30-35 GW of new electricity generation capacity within 20 years to plug the energy gap resulting from increased demand and the expected closure of existing power plants. It also states that, based on existing policies, renewable energy is likely to contribute around 5% of the UK's consumption by 2020, rather than the 20% target mentioned in the 2006 Energy Review.

Government Strategy

In summary, the government's proposed strategy involves 6 components:

- Establishing an international framework to tackle climate change, including the stabilisation of atmospheric greenhouse gas concentrations and a stronger European Union Emissions Trading Scheme

- Providing legally binding carbon targets for the whole UK economy, reducing emissions through the implementation of the Climate Change Bill.
- Making further progress in achieving fully competitive and transparent international markets, including further liberalisation of the European Union energy market.
- Encouraging more energy saving through better information, incentives and regulation
- Providing more support for low carbon technologies, including increased international and domestic public-private sector collaboration in the areas of research, development, demonstration and deployment – for example though the launch of the Energy Technologies Institute and the Environmental Transformation Fund.

To achieve the government's aims, the White Paper proposes a number of practical measures, including:

Energy Conservation

Businesses:

- A new mandatory cap and trade scheme for organisations consuming more than 6,000 MWh of electricity per year, to be known as the Carbon Reduction Commitment.
- The introduction of Energy Performance Certificates for business premises and Display Energy Certificates for public sector organisations.
- The extension of smart metring to most business premises within 5 years.

Homes:

- A requirement for all new homes to be zero-carbon buildings as soon as practically possible and preferably by 2016.
- Improving the energy efficiency of existing homes.
- Improving the efficiency of consumer electronics and domestic appliances, and the possible phase-out of inefficient light bulbs by around 2011.
- Increasing the Carbon Emission Reduction Target for the electricity and gas industries for 2008-2011.

- A requirement that new domestic electricity metres should have real time displays from 2008, and a commitment to upgrade existing domestic metres on request.

Transport:

- The introduction of a Low Carbon Transport Innovation Strategy
- Support for including aviation within the EU Emissions Trading Scheme

Energy Supply

- The introduction of a Biomass Strategy to expand the use of biomass as an energy source.
- Measures to grow distributed electricity generation and distributed heat generation alongside the centralised system.
- A reconfirmation that, under the Renewables Obligation, renewable energy should supply 10% of electricity generation by 2010, an 'aspiration' to achieve 20% by 2020, together with the introduction of bands within the Obligation to support different renewable technologies.
- The launch in November 2007 of a competition to demonstrate commercial-scale carbon capture and storage technology
- A 'preliminary view is that it is in the public interest to give the private sector the option of investing in new nuclear power stations'. A consultation on this was launched at the same time as the White Paper.
- The introduction of the Renewable Transport Fuel Obligation in 2008-2009, with a commitment that biofuels should provide 5% of transport fuel by 2010-2011.
- Measures to support the recovery of the remaining oil and gas reserves from the North Sea.
- Removing barriers to developing new energy infrastructure and power plants through reform of planning permission processes, as detailed in the 2007 Planning White Paper: *Planning for a Sustainable Future.*

Response of the Scottish Government

The Scottish Government responded to the UK government paper by making clear that it was against new nuclear power stations being built in Scotland and had the power to prevent any being built. In a statement to parliament, Energy Minister Jim Mather stated "Members will be aware that Greenpeace, backed by the courts, have forced the UK Government to consult properly on the future role of nuclear power. We will respond and we will make clear that we do not want and do not need new nuclear power in Scotland. If an application were to be submitted for a new nuclear power station that will be for Scottish Ministers to determine. We would be obliged to look at it - but given our policy position, our generating capacity, our multiplicity of energy sources and our strong alternative strategies such an application would be unlikely to find favour with this administration."

Status

Following a judicial review requested by Greenpeace, on February 15, 2007 elements of the 2006 Energy Review were ruled 'seriously flawed', and 'not merely inadequate but also misleading'. As a result, plans to build a new generation of nuclear power plants were ruled illegal.

Background

The UK Government published its White Paper on Energy ("Our Energy Future – creating a Low Carbon Economy") in 2003, establishing a formal energy policy for the UK for the first time in 20 years. Essentially, the White Paper recognised that a limitation of carbon dioxide (CO2 – the main gas contributing to global climate change) was going to be necessary. It committed the UK to working towards a 60% reduction in carbon dioxide emissions by 2050, and identified business opportunities in so doing: a recurrent theme throughout the document was "cleaner, smarter energy". It also claimed to be based on four pillars: the environment, energy reliability, affordable energy for the poorest and competitive markets.

However the White Paper focused more on analysing the issues than in providing detailed policy responses. Some detail began to filter through in a series of follow-on documents, including an Energy Efficiency Implementation Plan (April 2004) and the DTI Microgeneration Strategy "Our Energy Challenge" (March 2006). Nonetheless, most of the policies were a continuation of business as usual, with emphasis on market-led solutions and an expectation that consumers act rationally, for example in installing energy efficiency measures to make running cost savings.

However, in November 2005 it was announced that the Government, under DTI leadership, would undertake a full scale Energy Review, and over 500 organisations and individuals made detailed submissions as part of this review. Officially, the review was to take stock of the outcomes to date of the White Paper, which a particular focus on cutting carbon (emissions of which remained stubbornly high) and to look in more detail at security of supply, as the UK's oil and gas production from the North Sea had peaked, and Russia was seen as being a high-risk supplier of gas.

Unofficially, it was widely felt that the real reason behind the review was to allow nuclear power back into the energy debate, as it had been sidelined in the 2003 White Paper. That document had said "This white paper does not contain specific proposals for building new nuclear power stations. However we do not rule out the possibility that at some point in the future new nuclear build might be necessary if we are to meet our carbon targets. Before any decision to proceed with the building of new nuclear power stations, there will need to be the fullest public consultation and the publication of a further white paper setting out our proposals." The Energy Review was therefore to be this public consultation. A further White Paper was promised for early 2007.

Contents

The Energy Challenge: The Energy Review Report 2006

In the event, the Energy Review Report 2006 came out as a broader and more balanced document than critics (in advance) had expected. It started by reiterating the Government's four long-term goals for energy policy:

- To put the UK on a path to cut carbon dioxide emissions by some 60% by about 2050, with real progress by 2020;
- To maintain reliable energy supplies;
- To promote competitive markets in the UK and beyond, helping to raise the rate of sustainable economic growth and to improve productivity; and
- To ensure that every home is adequately and affordably heated.

It then identified two major long-term energy challenges:

- Tackling climate change, along with other nations, as global carbon emissions from human activity continue to grow; and
- Delivering secure, clean energy at affordable prices, as we become increasingly dependent on imports for our energy needs.

The Review took an internationalist response, stressing that the world's economies need to get on a path to being significantly less carbon-intensive, and noting rising global demand, especially from countries such as India and China. This means using less energy in products and services and changing the way energy is produced so that more of it comes from low-carbon sources. It also identified the need for a fairer distribution of energy around the world, and identified that many resources, especially of fossil fuels which are concentrated in just a few countries.

It placed its main concerns and proposals into three groups:

***Saving Energy*:** The starting point for reducing carbon emissions is to save energy. The challenge is to secure the heat, light and energy we need in homes and businesses in a way that cuts the amount of oil, gas and electricity used and the carbon dioxide emitted. Actions proposed include:

- Increasing information, e.g. through Home Information Packs
- Raising basic standards, removing inefficient goods from the market
- Making best use of the EU Emissions Trading Scheme and Climate Change Levy
- Making the Government estate carbon neutral by 2012
- Increasing the focus on energy efficient transport

***Cleaner Energy*:** Cost-effective ways of using less energy will help move towards the carbon reduction goal. But on their own they will not provide the solution to the challenges faced: there is also a need to make the energy used cleaner. Under this head, the Government considered:

- more distributed energy generation including low-carbon heat
- more use of community based systems, including CHP
- a strong commitment to carbon pricing in the UK, through improving the operation of the EU Emissions Trading Scheme
- a strengthened commitment to the Renewables Obligation
- proposals for reform of the planning regime for electricity projects
- a clear statement of our position on new nuclear build
- support for carbon capture and storage
- developing alternative fuels for transport

***The Energy Security Challenge*:** The challenges of reducing carbon emissions and ensuring security of supply are closely linked. Security of supply requires that we have good access to available fuel supplies, the infrastructure in place to transport them to centres of demand and effective markets so that supply meets demand in the most efficient way.

Many of the measures already described for tackling carbon emissions also contribute to the healthy diversity of energy sources that is necessary for meeting the energy security challenge.

There are two main security of supply challenges for the UK:

- Managing increased dependence on oil and gas imports, especially in the light of the global distribution of energy reserves and growing international demand; and
- Ensuring that the market delivers substantial and timely investment in electricity generating capacity and networks so that households and businesses have the electricity they need at affordable prices.

The Government's response is to continue to open up markets and to work internationally to develop strong relationships with suppliers, developing liberalised markets.

So where does nuclear power fit within this debate? Although it is mentioned a lot more in the Review compared to the White Paper (441 times, compared to 55 to be exact), the Government does not propose building new stations itself. Instead, it will leave it to the market, although it will ease some of the planning constraints (which it also aims to do for renewables) and look into providing a design authorisation procedure. However, as with many other aspects of the Energy Review Response, the document is not likely to be the last word on the subject, as there are plans for further consultation, and the establishment of further reviews and studies in issues such as identifying suitable sites, and managing the costs of decommissioning and long term waste management.

Issues Not Thoroughly Addressed by UK Policy

Despite some successes and stated goals, there are some issues that are incompletely addressed by UK policy. The principal such items are:

Peak Oil

In their October 2009 report, the Government-supported UK Energy Research Centre (UKERC) noted that the implications of reaching the peak of world oil production had, until the late 2000s, caused little concern among the world's governments. The UKERC report concluded that this peak could be expected before 2030, but that there is a 'significant risk' of a peak before 2020. The UK Government has no contingency plans for oil peaking before 2020.

The UKERC report's authors warn of the risk that 'rising oil prices will encourage the rapid development of carbon-intensive alternatives (such as coal) which will make it difficult or impossible to prevent dangerous climate change' and that 'early investment in low-carbon alternatives to conventional oil is of considerable importance' in avoiding this scenario. It is suggested that the current measures being established to address climate change may not be sufficient or rapid enough to address the challenges of peak oil, but that it will require 'both improved understanding and much greater awareness of the risks presented by global oil depletion' for further action to become politically feasible.

Other Issues

Loss of energy independence. Rather than creating an aggressive plan to lessen the impending loss of energy independence, the UK policy succumbs to that outcome, with resultant risks to future supplies and costs.

- Lack of strong national policy on transport fuel efficiency. While the city of London and other local councils have given incentives to hybrid vehicles, the national policy does not provide any real stimulus to highly fuel efficient vehicles. In fact, the government has done so little to inform the public about fuel efficient options that a survey released in March, 2006: "Some of the 1200 people surveyed... thought 'hybrid' meant two cars welded together. Others thought hybrids had to be plugged in at night."
- Lack of emphasis of energy conservation by reducing over-illumination, If the light level in a building is reduced then so is its energy usage.

Public Opinion

The UK results from the 1st Annual World Environment Review, published on June 5, 2007 revealed that:

- 81% are concerned about climate change.
- 79% think their Government should do more to tackle global warming.
- 73% think that the UK is too dependent on fossil fuels.
- 77% think that the UK is too reliant on foreign oil.
- 87% think that a minimum 25% of electricity should be generated from renewable energy sources.
- 24% think that the Government should do more to expand nuclear power.

- 56% are concerned about nuclear power.
- 76% are concerned about carbon dioxide emissions from developing countries.
- 61% think it appropriate for developed countries to demand restrictions on carbon dioxide emissions from developing countries.

Greenhouse Gas

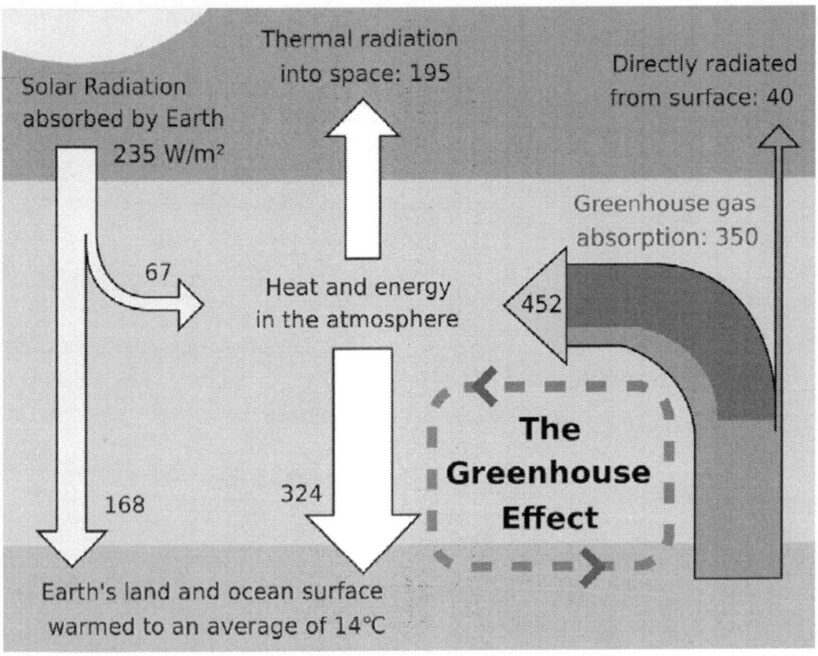

Figure: *Greenhouse effect schematic showing energy flows between space, the atmosphere, and Earth's surface. Energy influx and emittance are expressed in watts per square meter (W/m2).*

A greenhouse gas (sometimes abbreviated GHG) is a gas in an atmosphere that absorbs and emits radiation within the thermal infrared range. This process is the fundamental cause of the greenhouse effect. The primary greenhouse gases in the Earth's atmosphere are water vapour, carbon dioxide, methane, nitrous oxide, and ozone. In the Solar System, the atmospheres of Venus, Mars, and Titan also contain gases that cause greenhouse effects. Greenhouse gases greatly affect the temperature of the Earth; without them, Earth's surface would be on average about 33 °C (59 °F) colder than at present.

Since the beginning of the Industrial Revolution, the burning of fossil fuels has contributed to the increase in carbon dioxide in the atmosphere from 280 ppm to 390 ppm, despite the uptake of a large portion of the emissions through various natural "sinks" involved in the carbon cycle. Anthropogenic carbon dioxide (CO_2) emissions (i.e., emissions produced by human activities) come from combustion of carbonaceous fuels, principallywood, coal, oil, and natural gas.

Gases in Earth's Atmosphere

Greenhouse Gases

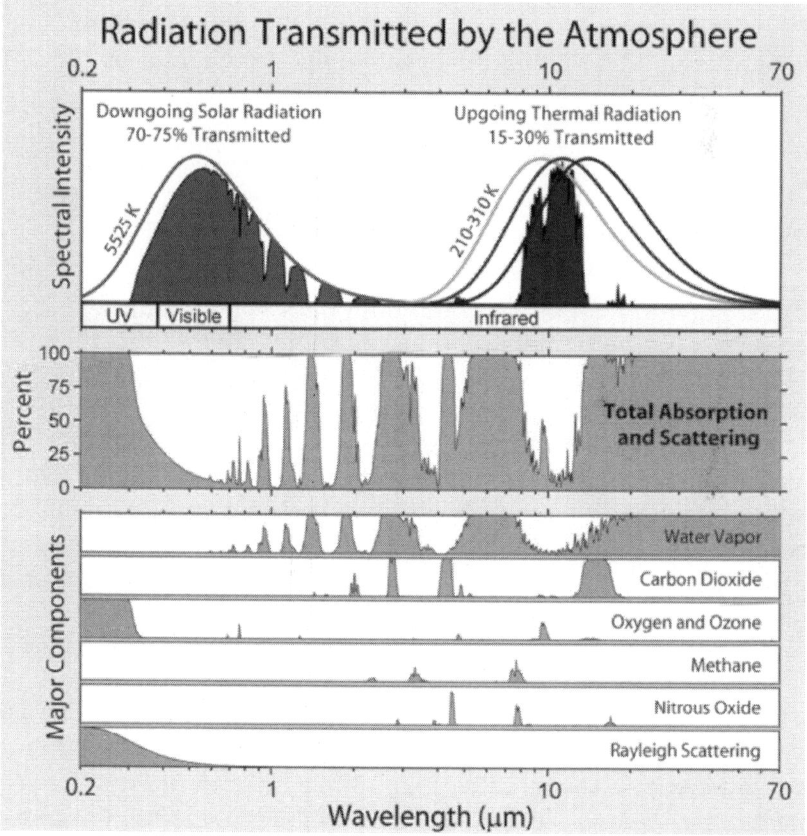

Figure : *Atmospheric absorption and scattering at different electromagnetic wavelengths. The largest absorption band of carbon dioxide is in the infrared.*

Greenhouse gases are those that can absorb and emit infrared radiation. In order, the most abundant greenhouse gases in Earth's atmosphere are:

- water vapour
- carbon dioxide
- methane

- nitrous oxide
- ozone.

Atmospheric concentrations of greenhouse gases are determined by the balance between sources (emissions of the gas from human activities and natural systems) and sinks (the removal of the gas from the atmosphere by conversion to a different chemical compound). The proportion of an emission remaining in the atmosphere after a specified time is the "Airborne fraction" (AF). More precisely, the annual AF is the ratio of the atmospheric increase in a given year to that year's total emissions. For CO_2 the AF over the last 50 years (1956–2006) has been increasing at $0.25 \pm 0.21\%$/year.

Non-greenhouse Gases

Although contributing to many other physical and chemical reactions, the major atmospheric constituents, nitrogen (N_2), oxygen (O_2), and argon (Ar), are not greenhouse gases. This is because molecules containing two atoms of the same element such as N_2 and O_2 and monatomic molecules such as Ar have no net change in their dipole moment when they vibrate and hence are almost totally unaffected by infrared light. Although molecules containing two atoms of different elements such as carbon monoxide (CO) or hydrogen chloride (HCl) absorb IR, these molecules are short-lived in the atmosphere owing to their reactivity and solubility. Because they do not contribute significantly to the greenhouse effect, they are usually omitted when discussing greenhouse gases.

Non-greenhouse Gases that have an Indirect Radiative Effect

Carbon monoxide has an indirect radiative effect by elevating concentrations of methane and tropospheric ozone through scavenging of atmospheric constituents (e.g., the hydroxyl radical, OH) that would otherwise destroy them. Carbon monoxide is created when carbon-containing fuels are burned incompletely. Through natural processes in the atmosphere, it is eventually oxidised to carbon dioxide. Carbon monoxide has an atmospheric lifetime of only a few months and as a consequence is spatially more variable than longer-lived gases.

Contribution of Clouds to Earth's Greenhouse Effect

The major non-gas contributor to the Earth's greenhouse effect, clouds, also absorb and emit infrared radiation and thus have an effect on radiative properties of the greenhouse gases. Clouds are water droplets or ice crystals suspended in the atmosphere.

Impact of a given Gas on the Overall Greenhouse Effect

Each gases' contribution to the greenhouse effect is affected by the characteristics of the gas, its abundance, and any indirect effects it may cause. For example, on a molecule-for-molecule basis the direct radiative effects of methane is about 72 times stronger greenhouse gas than carbon dioxide over a 20 year time frame but it is present in much smaller concentrations so that its total direct radiative effect is smaller. On the other hand, in addition to its direct radiative impact methane has a large indirect radiative effect because it contributes to ozone formation. Shindell *et al.* (2005) argue that the contribution to climate change from methane is at least double previous estimates as a result of this effect.

When these gases are ranked by their direct contribution to the greenhouse effect, the most important are:

Gas	Formula	Contribution(%)
Water vapour	H_2O	*36 – 72 %*
Carbon dioxide	CO_2	*9 – 26 %*
Methane	CH_4	*4 – 9 %*
Ozone	O_3	*3 – 7 %*

In addition to the main greenhouse gases listed above, other greenhouse gases include sulphur hexafluoride, hydrofluorocarbons and perfluorocarbons. Some greenhouse gases are not often listed. For example, nitrogen trifluoride has a high global warming potential (GWP) but is only present in very small quantities.

Proportion of Direct Effects at a Given Moment

It is not possible to state that a certain gas causes an exact percentage of the greenhouse effect. This is because some of the gases absorb and emit radiation at the same frequencies as others, so that the total greenhouse effect is not simply the sum of the influence of each gas. The higher ends of the ranges quoted are for each gas alone; the lower ends account for overlaps with the other gases. In addition, some gases such as methane are known to have large indirect effects that are still being quantified.

Atmospheric Lifetime

Aside from water vapour, which has a residence time of about nine days, major greenhouse gases are well-mixed, and take many years to leave the atmosphere. Although it is not easy to know with precision how long it takes greenhouse gases to leave the atmosphere, there are estimates for the principal greenhouse gases. Jacob (1999) defines the lifetime τ of an atmospheric species X in a one-box model as the average time that a molecule of X remains in the box. Mathematically can be defined as the ratio of the mass (in kg) of X in the box to its removal rate, which is the sum of the flow of X out of the box (), chemical loss of X (), and deposition of X () (all in kg/sec):. If one stopped pouring any of this gas into the box, then after a time , its concentration would be about halved.

The atmospheric lifetime of a species therefore measures the time required to restore equilibrium following a sudden increase or decrease in its concentration in the atmosphere. Individual atoms or molecules may be lost or deposited to sinks such as the soil, the oceans and other waters, or vegetation and other biological systems, reducing the excess to background concentrations. The average time taken to achieve this is the mean lifetime.

Carbon dioxide has a variable atmospheric lifetime, and cannot be specified precisely. The atmospheric lifetime of CO_2 is estimated of the order of 30–95 years. However, this excludes the balancing fluxes of CO_2 into the atmosphere from the geological reservoirs, which have slower characteristic rates. While more than half of the CO_2 emitted is currently removed from the atmosphere within a century, some fraction (about 20%) of emitted CO_2 remains in the atmosphere for many thousands of years.

Global Warming Potential

The global warming potential (GWP) depends on both the efficiency of the molecule as a greenhouse gas and its atmospheric lifetime. GWP is measured relative to the same mass of CO_2 and evaluated for a specific timescale. Thus, if a gas has a high radiative forcing but also a short lifetime, it will have a large GWP on a 20 year scale but a small one on a 100 year scale. Conversely, if a molecule has a longer atmospheric lifetime than CO_2 its GWP will increase with the timescale considered. Carbon dioxide is defined to have a GWP of 1 over all time periods.

Methane has an atmospheric lifetime of 12 ± 3 years and a GWP of 72 over 20 years, 25 over 100 years and 7.6 over 500 years. The decrease in GWP at longer times is because methane is degraded to water and CO_2 through chemical reactions in the atmosphere.

Natural and Anthropogenic Sources

Aside from purely human-produced synthetic halocarbons, most greenhouse gases have both natural and human-caused sources. During the pre-industrial Holocene, concentrations of existing gases were roughly constant. In the industrial era, human activities have added greenhouse gases to the atmosphere, mainly through the burning of fossil fuels and clearing of forests.

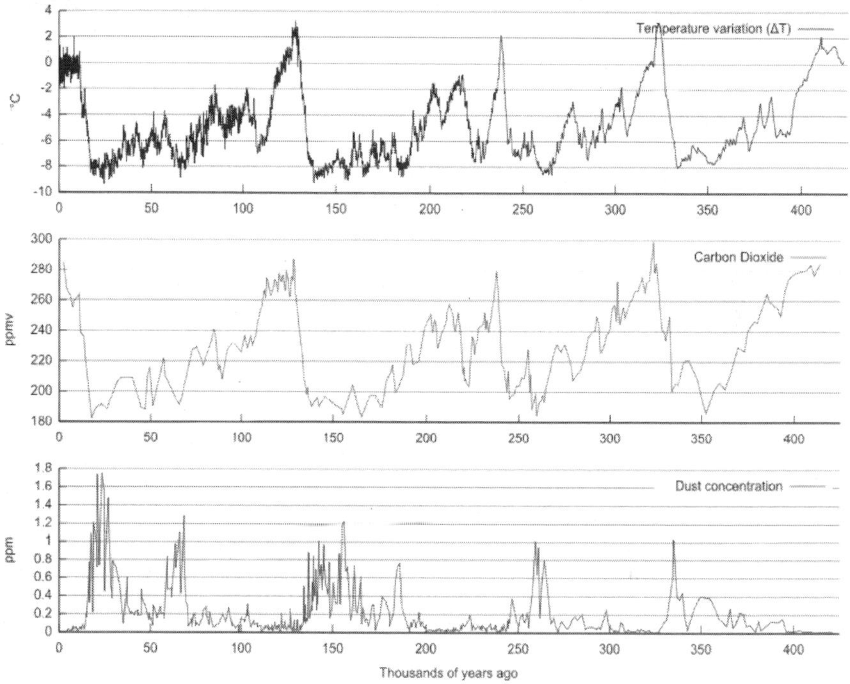

***Figure:** 400,000 years of ice core data.*

The 2007 Fourth Assessment Report compiled by the IPCC (AR4) noted that "changes in atmospheric concentrations of greenhouse gases and aerosols, land cover and solar radiation alter the energy balance of the climate system", and concluded that "increases in anthropogenic greenhouse gas concentrations is very likely to have caused most of the increases in global average temperatures since the mid-20th century". In AR4, "most of" is defined as more than 50%.

Gas	*Preindustrial level*	*Current level*	*Increase since 1750*	*Radiative forcing (W/m²)*
Carbon dioxide	280 ppm	393 ppm	113 ppm	1.46
Methane	700 ppb	1745 ppb	1045 ppb	0.48
Nitrous oxide	270 ppb	314 ppb	44 ppb	0.15
CFC-12	0	533 ppt	533 ppt	0.17

Ice cores provide evidence for variation in greenhouse gas concentrations over the past 800,000 years. Both CO_2 and CH_4 vary between glacial and interglacial phases, and concentrations of these gases correlate strongly with temperature. Direct data does not exist for periods earlier than those represented in the ice core record, a record which indicates CO_2 mole fractions staying within a range of between 180 ppm and 280 ppm throughout the last 800,000 years, until the increase of the last 250 years. However, various proxies and modelling suggests larger variations in past epochs; 500 million years ago CO_2 levels were likely 10 times higher than now. Indeed higher CO2 concentrations are thought to have prevailed throughout most of the Phanerozoic eon, with concentrations four to six times current concentrations during the Mesozoic era, and ten to fifteen times current concentrations during the early Palaeozoic era until the middle of the Devonian period, about 400 Ma. The spread of land plants is thought to have reduced CO_2 concentrations during the late Devonian, and plant activities as both sources and sinks of CO_2 have since been important in providing stabilising feedbacks. Earlier still, a 200-million year period of intermittent, widespread glaciation extending close to the equator (Snowball Earth) appears to have been ended suddenly, about 550 Ma, by a colossal volcanic outgassing which raised the CO_2 concentration of the atmosphere abruptly to 12%, about 350 times modern levels, causing extreme greenhouse conditions and carbonate deposition as limestone at the rate of about 1 mm per day. This episode marked the close of the Precambrian eon, and was succeeded by the generally warmer conditions of the Phanerozoic, during which multicellular animal and plant life evolved. No volcanic carbon dioxide emission of comparable scale has occurred since. In the modern era, emissions to the atmosphere from volcanoes are only about 1% of emissions from human sources.

Anthropogenic Greenhouse Gases

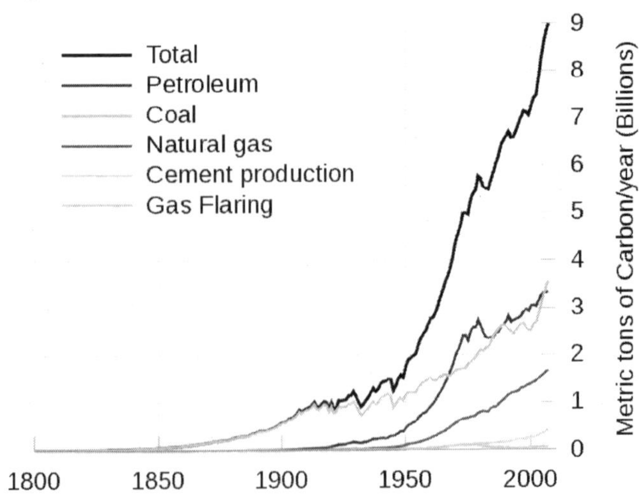

Figure : *Modern global anthropogenic carbonemissions.*

Since about 1750 human activity has increased the concentration of carbon dioxide and other greenhouse gases. Measured atmospheric concentrations of carbon dioxide are currently 100 ppm higher than pre-industrial levels. Natural sources of carbon dioxide are more than 20 times greater than sources due to human activity, but over periods longer than a few years natural sources are closely balanced by natural sinks, mainly photosynthesis of carbon compounds by plants and marine plankton. As a result of this balance, the atmospheric mole fraction of carbon dioxide remained between 260 and 280 parts per million for the 10,000 years between the end of the last glacial maximum and the start of the industrial era.

It is likely that anthropogenic warming, such as that due to elevated greenhouse gas levels, has had a discernible influence on many physical and biological systems. Warming is projected to affect various issues such as freshwater resources, industry, food and health.

The Main Sources of Greenhouse Gases Due to Human Activity are: burning of fossil fuels and deforestation leading to higher carbon dioxide concentrations in the air.

Land use change (mainly deforestation in the tropics) account for up to one third of total anthropogenic CO_2 emissions livestock enteric fermentation and manure management, paddy rice farming, land use and wetland changes, pipeline losses, and covered vented landfill emissions leading to higher methane atmospheric concentrations. Many of the newer style fully vented septic systems that enhance and target the fermentation process also are sources of atmospheric methane.

Use of chlorofluorocarbons (CFCs) in refrigeration systems, and use of CFCs and halons in fire suppression systems and manufacturing processes.

Agricultural activities, including the use of fertilizers, that lead to higher nitrous oxide (N_2O) concentrations.

The US Environmental Protection Agency (EPA) ranks the major greenhouse gas contributing end-user sectors in the following order: industrial, transportation, residential, commercial and agricultural. Major sources of an individual's greenhouse gas include home heating and cooling, electricity consumption, and transportation.

Corresponding conservation measures are improving home building insulation, installing geothermal heat pumps and compact fluorescent lamps, and choosing energy-efficient vehicles.

The seven sources of CO_2 from fossil fuel combustion are (with percentage contributions for 2000–2004):

Seven main fossil fuel combustion sources	***Contribution (%)***
Liquid fuels (e.g., gasoline, fuel oil)	36 %
Solid fuels (e.g., coal)	35 %
Gaseous fuels (e.g., natural gas)	20 %
Cement production	3 %
Flaring gas industrially and at wells	< 1 %
Non-fuel hydrocarbons	< 1 %
"International bunker fuels" of transport not included in national inventories	4 %

Carbon dioxide, methane, nitrous oxide and three groups of fluorinated gases (sulphur hexafluoride, HFCs, and PFCs) are the major greenhouse gases and the subject of the Kyoto Protocol, which came into force in 2005.

Although CFCs are greenhouse gases, they are regulated by the Montreal Protocol, which was motivated by CFCs' contribution to ozone depletion rather than by their contribution to global warming. Note that ozone depletion has only a minor role in greenhouse warming though the two processes often are confused in the media.

On December 7, 2009, the US Environmental Protection Agency released its final findings on greenhouse gases, declaring that "greenhouse gases (GHGs) threaten the public health and welfare of the American people". The finding applied to the same "six key well-mixed greenhouse gases" named in the Kyoto Protocol: carbon dioxide, methane, nitrous oxide, hydrofluorocarbons, perfluorocarbons, and sulphur hexafluoride.

Role of Water Vapour

Water vapour accounts for the largest percentage of the greenhouse effect, between 36% and 66% for clear sky conditions and between 66% and 85% when including clouds. Water vapour concentrations fluctuate regionally, but human activity does not significantly affect water vapour concentrations except at local scales, such as near irrigated fields. The atmospheric concentration of vapour is highly variable, from less than 0.01% in extremely cold regions up to 20% in warm, humid regions.

The average residence time of a water molecule in the atmosphere is only about nine days, compared to years or centuries for other greenhouse gases such as CH_4 and CO_2. Thus, water vapour responds to and amplifies effects of the other greenhouse gases. The Clausius-Clapeyron relation establishes that air can hold more water vapour per unit volume when it warms.

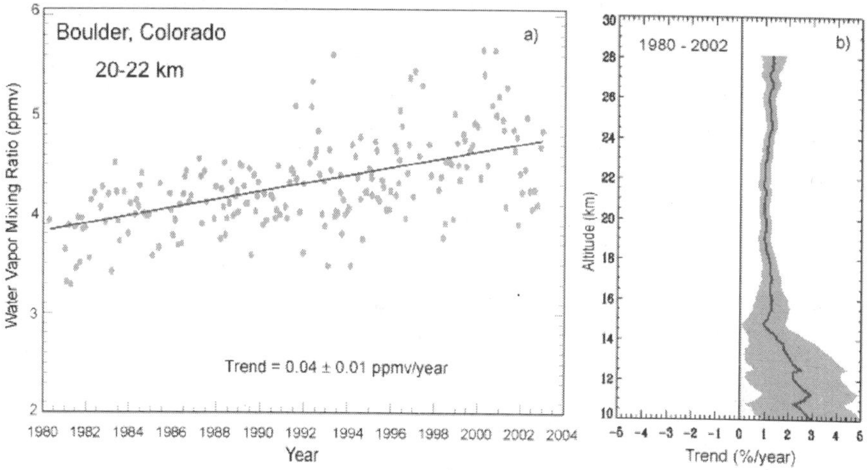

Figure : *Increasing water vapour in the stratosphere at Boulder, Colorado.*

This and other basic principles indicate that warming associated with increased concentrations of the other greenhouse gases also will increase the concentration of water vapour. Because water vapour is a greenhouse gas, this results in further warming and so is a "positive feedback" that amplifies the original warming. Eventually other earth processes offset these positive feedbacks, stabilising the global temperature at a new equilibrium and preventing the loss of earth's water through a Venus-like runaway greenhouse effect.

Atmospheric Concentration

Measurements from Antarctic ice cores show that before industrial emissions started atmospheric CO_2 mole fractions were about 280 parts per million (ppm), and stayed between 260 and 280 during the preceding ten thousand years. Carbon dioxide mole fractions in the atmosphere have gone up by approximately 35 percent since the 1900s, rising from 280 parts per million by volume to 387 parts per million in 2009. One study using evidence from stomata of fossilised leaves suggests greater variability, with carbon dioxide mole fractions above 300 ppm during the period seven to ten thousand years ago, though others have argued that these findings more likely reflect calibration or contamination problems rather than actual CO_2 variability. Because of the way air is trapped in ice (pores in the ice close off slowly to form bubbles deep within the firn) and the time period represented in each ice sample analysed, these figures represent averages of atmospheric concentrations of up to a few centuries rather than annual or decadal levels.

Since the beginning of the Industrial Revolution, the concentrations of most of the greenhouse gases have increased. For example, the mole fraction of carbon dioxide has increased from 280 ppm by about 36% to 380 ppm, or 100 ppm over modern pre-industrial levels. The first 50 ppm increase took place in about 200 years, from the start of the Industrial Revolution to around 1973; however the next 50 ppm increase took place in about 33 years, from 1973 to 2006.

Recent data also shows that the concentration is increasing at a higher rate. In the 1960s, the average annual increase was only 37% of what it was in 2000 through 2007.

Today, the stock of carbon in the atmosphere increases by more than 3 million tonnes per annum (0.04%) compared with the existing stock. This increase is the result of human activities by burning fossil fuels, deforestation and forest degradation in tropical and boreal regions.

The other greenhouse gases produced from human activity show similar increases in both amount and rate of increase. Many observations are available online in a variety of Atmospheric Chemistry Observational Databases.

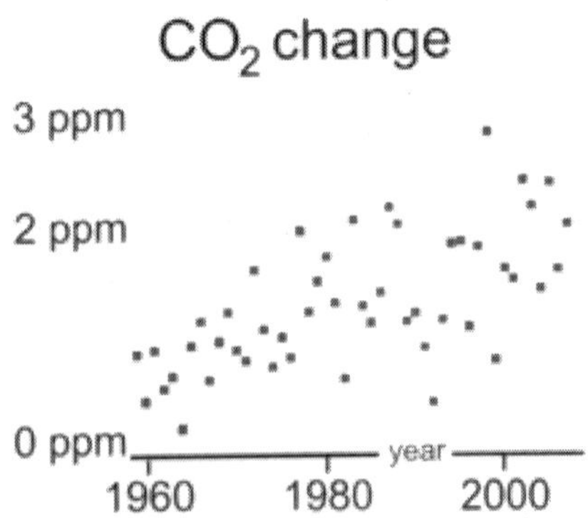

***Figure :** Recent year-to-year increase of atmospheric CO_2.*

Between the period 1970 to 2004, GHG emissions (measured in CO_2-equivalent) increased at an average rate of 1.6% per year, with CO_2 emissions from the use of fossil fuels growing at a rate of 1.9% per year. Total anthropogenic emissions at the end of 2009 were estimated at 49.5 gigatonnes CO_2-equivalent. These emissions include CO_2 from fossil fuel use and from land use, as well as emissions of methane, nitrous oxide and other GHGs covered by the Kyoto Protocol.

At present, the two primary sources of CO_2 emissions are from burning coal used for electricity generation and petroleum used for motor transport.

Regional and National Attribution of Emissions

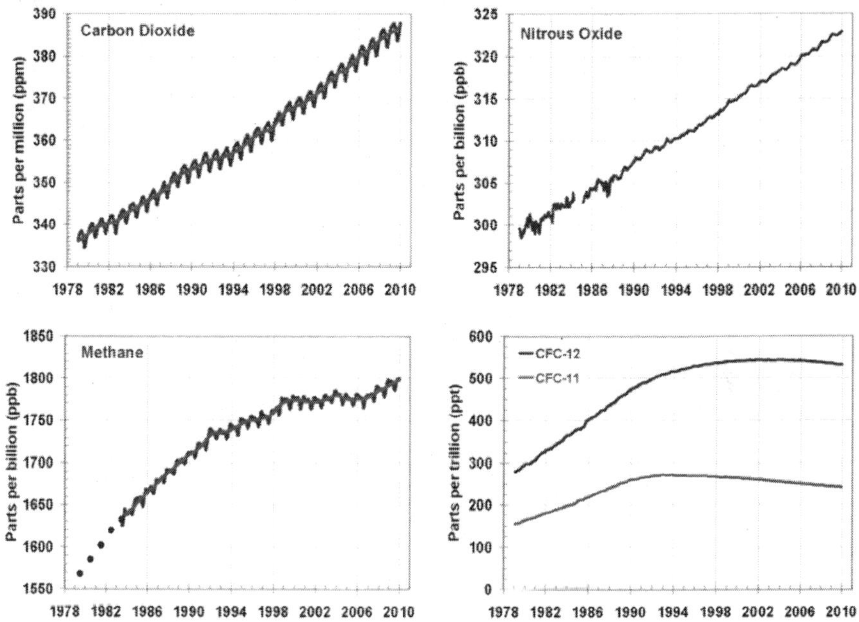

Figure : *Major greenhouse gas trends.*

There are several different ways of measuring GHG emissions.

Some variables that have been reported include:

Definition of measurement boundaries. Emissions can be attributed geographically, to the area where they were emitted (the territory principle) or by the activity principle to the territory that caused the emissions to be produced. These two principles would result in different totals when measuring for example the importation of electricity from one country to another or the emissions at an international airport. The time horizon of different GHGs. Contribution of a given GHG is reported as a CO_2 equivalent; the calculation to determine this takes into account how long that gas remains in the atmosphere. This is not always known accurately and calculations must be regularly updated to take into account new information.

What sectors are included in the calculation (e.g. energy industries, industrial processes, agriculture etc.). There is often a conflict between transparency and availability of data.

The measurement protocol itself. This may be via direct measurement or estimation; the four main methods are the emission factor-based method, the mass balance method, the predictive emissions monitoring system and the continuing emissions monitoring systems. The methods differ in accuracy, but also in cost and usability.

The different measures are sometimes used by different countries in asserting various policy/ethical positions to do with climate change. This use of different measures leads to a lack of comparability, which is problematic when monitoring progress towards targets. There are arguments for the adoption of a common measurement tool, or at least the development of communication between different tools.

Emissions may be measured over long time periods. This measurement type is called historical or cumulative emissions. Cumulative emissions give some indication of who is responsible for the build-up in the atmospheric concentration of GHGs.

The national accounts balance would be positively related to carbon emissions. The national accounts balance shows the difference between exports and imports. For many richer nations, such as the United States, the accounts balance is negative because more goods are imported than they are exported. This is mostly due to the fact that it is cheaper to produce goods outside of developed countries, leading the economies of developed countries to become increasingly dependent on services and not goods. We believed that a positive accounts balance would means that more production was occurring in a country, so more factories working would increase carbon emission levels.

Emissions may also be measured across shorter time periods. Emissions changes may, for example, be measured against a base year of 1990. 1990 was used in the United Nations Framework Convention on Climate Change (UNFCCC) as the base year for emissions, and is also used in the Kyoto Protocol (some gases are also measured from the year 1995). A country's emissions may also be reported as a proportion of global emissions for a particular year. Another measurement is of per capita emissions. This divides a country's total annual emissions by its mid-year population. Per capita emissions may be based on historical or annual emissions.

Greenhouse Gas Intensity and Land-use Change

Herzog *et al.* (2006) defined greenhouse gas intensity as GHG emissions divided by economic output. GHG intensities are subject to uncertainty over whether they are calculated using market exchange rates (MER) or purchasing power parity (PPP). Calculations based on MER suggest large differences in intensities between developed and developing countries, whereas calculations based on PPP show smaller differences.

Land-use change, e.g., the clearing of forests for agricultural use, can affect the concentration of GHGs in the atmosphere by altering how much carbon flows out of the atmosphere into carbon sinks. Accounting for land-use change can be understood as an attempt to measure "net" emissions, i.e., gross emissions from all GHG sources minus the removal of emissions from the atmosphere by carbon sinks.

There are substantial uncertainties in the measurement of net carbon emissions. Additionally, there is controversy over how carbon sinks should be allocated between different regions and over time. For instance, concentrating on more recent changes in carbon sinks is likely to favour those regions that have deforested earlier, e.g., Europe.

Cumulative and Historical Emissions

Cumulative anthropogenic (i.e., human-emitted) emissions of CO_2 from fossil fuel use are a major cause of global warming, and give some indication of which countries have contributed most to human-induced climate change.

Top-5 historic CO_2 contributors by region over the years 1800 to 1988 (in %)		
Region	***Industrial CO_2***	***Total CO_2***
OECD North America	33.2	29.7
OECD Europe	26.1	16.6
Former USSR	14.1	12.5
China	5.5	6.0
Eastern Europe	5.5	4.8

Overall, developed countries accounted for 83.8% of industrial CO_2 emissions over this time period, and 67.8% of total CO_2 emissions. Developing countries accounted for industrial CO_2 emissions of 16.2% over this time period, and 32.2% of total CO_2 emissions. The estimate of total CO_2 emissions includes biotic carbon emissions, mainly from deforestation. Banuri *et al.* (1996) calculated per capita cumulative emissions based on then-current population. The ratio in per capita emissions between industrialised countries and developing countries was estimated to be more than 10 to 1.

Including biotic emissions brings about the same controversy mentioned earlier regarding carbon sinks and land-use change (Banuri *et al.*, 1996). The actual calculation of net emissions is very complex, and is affected by how carbon sinks are allocated between regions and the dynamics of the climate system.

Non-OECD countries accounted for 42% of cumulative energy-related CO_2 emissions between 1890-2007. Over this time period, the

US accounted for 28% of emissions; the EU, 23%; Russia, 11%; China, 9%; other OECD countries, 5%; Japan, 4%; India, 3%; and the rest of the world, 18%.

Changes Since a Particular Base Year

Between 1970-2004, global growth in annual CO_2 emissions was driven by North America, Asia, and the Middle East. The sharp acceleration in CO_2 emissions since 2000 to more than a 3% increase per year (more than 2 ppm per year) from 1.1% per year during the 1990s is attributable to the lapse of formerly declining trends in carbon intensity of both developing and developed nations. China was responsible for most of global growth in emissions during this period. Localised plummeting emissions associated with the collapse of the Soviet Union have been followed by slow emissions growth in this region due to more efficient energy use, made necessary by the increasing proportion of it that is exported. In comparison, methane has not increased appreciably, and N_2O by 0.25% y^{-1}.

Using different base years for measuring emissions has an effect on estimates of national contributions to global warming. This can be calculated by dividing a country's highest contribution to global warming starting from a particular base year, by that country's minimum contribution to global warming starting from a particular base year.

Choosing between different base years of 1750, 1900, 1950, and 1990 has a significant effect for most countries. Within the G8 group of countries, it is most significant for the UK, France and Germany. These countries have a long history of CO_2 emissions.

Annual Emissions

Annual per capita emissions in the industrialised countries are typically as much as ten times the average in developing countries. Due to China's fast economic development, its annual per capita emissions are quickly approaching the levels of those in the Annex I group of the Kyoto Protocol (i.e., the developed countries excluding the USA). Other countries with fast growing emissions are South Korea, Iran, and Australia. On the other hand, annual per capita emissions of the EU-15 and the USA are gradually decreasing over time. Emissions in Russia and the Ukraine have decreased fastest since 1990 due to economic restructuring in these countries.

Energy statistics for fast growing economies are less accurate than those for the industrialised countries. For China's annual emissions in

2008, the Netherlands Environmental Assessment Agency estimated an uncertainty range of about 10%.

Embedded Emissions

One way of attributing greenhouse gas (GHG) emissions is to measure the embedded emissions (also referred to as "embodied emissions") of goods that are being consumed. Emissions are usually measured according to production, rather than consumption. For example, in the main international treaty on climate change (the UNFCCC), countries report on emissions produced within their borders, e.g., the emissions produced from burning fossil fuels. Under a production-based accounting of emissions, embedded emissions on imported goods are attributed to the exporting, rather than the importing, country. Under a consumption-based accounting of emissions, embedded emissions on imported goods are attributed to the importing country, rather than the exporting, country.

Davis and Caldeira (2010) found that a substantial proportion of CO_2 emissions are traded internationally. The net effect of trade was to export emissions from China and other emerging markets to consumers in the US, Japan, and Western Europe. Based on annual emissions data from the year 2004, and on a per-capita consumption basis, the top-5 emitting countries were found to be (in CO_2 per person, per year): Luxembourg (34.7), the US (22.0), Singapore (20.2), Australia (16.7), and Canada (16.6).

Effect of Policy

Rogner *et al.* (2007) assessed the effectiveness of policies to reduce emissions (mitigation of climate change). They concluded that mitigation policies undertaken by UNFCCC Parties were inadequate to reverse the trend of increasing GHG emissions. The impacts of population growth, economic development, technological investment, and consumption had overwhelmed improvements in energy intensities and efforts to decarbonise (energy intensity is a country's total primary energy supply (TPES) per unit of GDP (Rogner *et al.*, 2007). TPES is a measure of commercial energy consumption (World Bank2010)).

Projections

Based on then-current energy policies, Rogner et al. (2007) projected that energy-related CO_2 emissions in 2030 would be 40-110% higher than in 2000. Two-thirds of this increase was projected to come from non-Annex I countries. Per capita emissions in Annex I countries were still projected to remain substantially higher than per capita emissions in non-Annex I countries. Projections consistently showed a

25-90% increase in the Kyoto gases (carbon dioxide, methane, nitrous oxide, sulphur hexafluoride) compared to 2000.

IEA (2007) estimated future cumulative energy-related CO_2 emissions for several countries. Their reference scenario projected cumulative energy-related CO_2 emissions between the years 1900 and 2030. In this scenario, China's share of cumulative emissions rises to 16%, approaching that of the United States (25%) and the European Union (18%). India's cumulative emissions (4%) approach those of Japan (4%).

Relative CO_2 Emission from Various Fuels

One litre of gasoline, when used as a fuel, produces 2.32 kg (1.3 cubic metres) of carbon dioxide, a greenhouse gas. One US gallon produces 19.4 lb (172.65 cubic feet)

Removal from the Atmosphere ("Sinks")

Natural Processes

Greenhouse gases can be removed from the atmosphere by various processes, as a consequence of:

- a physical change (condensation and precipitation remove water vapour from the atmosphere).
- a chemical reaction within the atmosphere. For example, methane is oxidised by reaction with naturally occurring hydroxyl radical, OH · and degraded to CO_2 and water vapour (CO_2 from the oxidation of methane is not included in the methane Global warming potential). Other chemical reactions include solution and solid phase chemistry occurring in atmospheric aerosols.
- a physical exchange between the atmosphere and the other compartments of the planet. An example is the mixing of atmospheric gases into the oceans.
- a chemical change at the interface between the atmosphere and the other compartments of the planet. This is the case for CO_2, which is reduced by photosynthesis of plants, and which, after dissolving in the oceans, reacts to form carbonic acid and bicarbonate and carbonate ions.
- a photochemical change. Halocarbons are dissociated by UV light releasing Cl · and F · as free radicals in the stratosphere with harmful effects on ozone (halocarbons are generally too stable to disappear by chemical reaction in the atmosphere).

Negative Emissions

Bio-energy with carbon capture and storage, Carbon dioxide air capture, Geoengineering, and Greenhouse gas remediation

There are a number of technologies that remove emissions of greenhouse gases from the atmosphere. Most widely analysed are those which remove carbon dioxide from the atmosphere, either to geologic formations such as bio-energy with carbon capture and storage and carbon dioxide air capture, or to the soil as in the case with biochar. It has been pointed out by the IPCC, that many long-term climate scenario models require large scale manmade negative emissions in order to avoid serious climate change.

History of Scientific Research

Late 19th century scientists experimentally discovered that N_2 and O_2 do not absorb infrared radiation (called, at that time, "dark radiation") while, on the contrary, water, both as true vapour and condensed in the form of microscopic droplets suspended in clouds, as well as CO_2 and other poly-atomic gaseous molecules, do absorb infrared radiation. It was recognised in the early 20th century that the greenhouse gases in the atmosphere caused the Earth's overall temperature to be higher than it would be without them. During the late 20th century, a scientific consensus evolved that increasing concentrations of greenhouse gases in the atmosphere are causing a substantial rise in global temperatures and changes to other parts of the climate system, with consequences for the environment and for human health.

Renewable Energy

Renewable energy is energy which comes from natural resources such as sunlight, wind, rain, tides, and geothermal heat, which are renewable (naturally replenished). About 16% of global final energy consumption comes from renewables, with 10% coming from traditional biomass, which is mainly used for heating, and 3.4% from hydroelectricity. New renewables (small hydro, modern biomass, wind, solar, geothermal, and biofuels) accounted for another 3% and are growing very rapidly. The share of renewables in electricity generation is around 19%, with 16% of global electricity coming from hydroelectricity and 3% from new renewables.

Wind power is growing at the rate of 30% annually, with a worldwide installed capacity of 238,000 megawatts (MW) at the end of

2011, and is widely used in Europe, Asia, and the United States. At the end of 2011 the photovoltaic (PV) capacity worldwide was 67,000 MW, and PV power stations are popular in Germany and Italy. Solar thermal power stations operate in the USA and Spain, and the largest of these is the 354 megawatt (MW) SEGS power plant in the Mojave Desert. The world's largest geothermal power installation is the Geysers in California, with a rated capacity of 750 MW.

Brazil has one of the largest renewable energy programs in the world, involving production of ethanol fuel from sugarcane, and ethanol now provides 18% of the country's automotive fuel. Ethanol fuel is also widely available in the USA.

While many renewable energy projects are large-scale, renewable technologies are also suited to rural and remote areas, where energy is often crucial in human development. As of 2011, small solar PV systems provide electricity to a few million households, and micro-hydro configured into mini-grids serves many more. Over 44 million households use biogas made in household-scale digesters for lighting and/or cooking, and more than 166 million households rely on a new generation of more-efficient biomass cookstoves. United Nations' Secretary-General Ban Ki-moon has said that renewable energy has the ability to lift the poorest nations to new levels of prosperity.

Climate change concerns, coupled with high oil prices, peak oil, and increasing government support, are driving increasing renewable energy legislation, incentives and commercialisation. New government spending, regulation and policies helped the industry weather the global financial crisis better than many other sectors. According to a 2011 projection by the International Energy Agency, solar power generators may produce most of the world's electricity within 50 years, dramatically reducing the emissions of greenhouse gases that harm the environment.

Overview

Renewable energy flows involve natural phenomena such as sunlight, wind, tides, plant growth, and geothermal heat, as the International Energy Agency explains:

Renewable energy is derived from natural processes that are replenished constantly. In its various forms, it derives directly from the sun, or from heat generated deep within the earth. Included in the definition is electricity and heat generated from solar, wind, ocean, hydropower, biomass, geothermal resources, and biofuels and hydrogen derived from renewable resources.

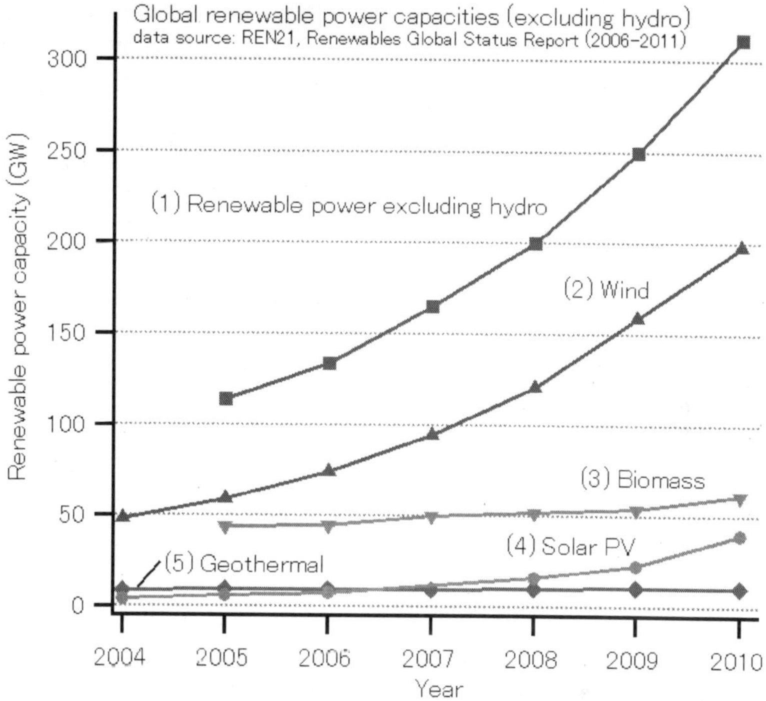

Figure: *Global renewable power capacity excluding hydro*

Renewable energy replaces conventional fuels in four distinct areas: electricity generation, hot water/ space heating, motor fuels, and rural (off-grid) energy services:

- Power generation. Renewable energy provides 19% of electricity generation worldwide. Renewable power generators are spread across many countries, and wind power alone already provides a significant share of electricity in some areas: for example, 14% in the U.S. state of Iowa, 40% in the northern German state of Schleswig-Holstein, and 20% in Denmark. Some countries get most of their power from renewables, including Iceland and Paraguay (100%), Norway (98%), Brazil (86%), Austria (62%), New Zealand (65%), and Sweden (54%).
- Heating. Solar hot water makes an important contribution to renewable heat in many countries, most notably in China, which now has 70% of the global total (180 GWth). Most of these systems are installed on multi-family apartment buildings and meet a portion of the hot water needs of an estimated 50–60 million households in China. Worldwide, total installed solar water heating systems meet a portion of the water heating needs

of over 70 million households. The use of biomass for heating continues to grow as well. In Sweden, national use of biomass energy has surpassed that of oil. Direct geothermal for heating is also growing rapidly.

- Transport fuels. Renewable biofuels have contributed to a significant decline in oil consumption in the United States since 2006. The 93 billion litres of biofuels produced worldwide in 2009 displaced the equivalent of an estimated 68 billion litres of gasoline, equal to about 5% of world gasoline production.

Mainstream Forms of Renewable Energy

Wind Power

Airflows can be used to run wind turbines. Modern wind turbines range from around 600 kW to 5 MW of rated power, although turbines with rated output of 1.5–3 MW have become the most common for commercial use; the power output of a turbine is a function of the cube of the wind speed, so as wind speed increases, power output increases dramatically. Areas where winds are stronger and more constant, such as offshore and high altitude sites, are preferred locations for wind farms. Typical capacity factors are 20-40%, with values at the upper end of the range in particularly favourable sites.

Globally, the long-term technical potential of wind energy is believed to be five times total current global energy production, or 40 times current electricity demand. This could require wind turbines to be installed over large areas, particularly in areas of higher wind resources. Offshore resources experience average wind speeds of ~90% greater than that of land, so offshore resources could contribute substantially more energy.

Hydropower

Energy in water can be harnessed and used. Since water is about 800 times denser than air, even a slow flowing stream of water, or moderate sea swell, can yield considerable amounts of energy. There are many forms of water energy:

- Hydroelectric energy is a term usually reserved for large-scale hydroelectric dams. Examples are the Grand Coulee Dam in Washington State and the Akosombo Dam in Ghana.
- Micro hydro systems are hydroelectric power installations that typically produce up to 100 kW of power. They are often used in water rich areas as a remote-area power supply (RAPS).

Run-of-the-river hydroelectricity systems derive kinetic energy from rivers and oceans without using a dam.

Solar Energy

Solar energy is the energy derived from the sun through the form of solar radiation. Solar powered electrical generation relies on photovoltaics and heat engines. A partial list of other solar applications includes space heating and cooling through solar architecture, daylighting, solar hot water, solar cooking, and high temperature process heat for industrial purposes.

Solar technologies are broadly characterised as either passive solar or active solar depending on the way they capture, convert and distribute solar energy. Active solar techniques include the use of photovoltaic panels and solar thermal collectors to harness the energy. Passive solar techniques include orienting a building to the Sun, selecting materials with favourable thermal mass or light dispersing properties, and designing spaces that naturally circulate air.

Biomass

Biomass (plant material) is a renewable energy source because the energy it contains comes from the sun. Through the process of photosynthesis, plants capture the sun's energy. When the plants are burnt, they release the sun's energy they contain. In this way, biomass functions as a sort of natural battery for storing solar energy. As long as biomass is produced sustainably, with only as much used as is grown, the battery will last indefinitely.

In general there are two main approaches to using plants for energy production: growing plants specifically for energy use (known as first and third-generation biomass), and using the residues (known as second-generation biomass) from plants that are used for other things. The best approaches vary from region to region according to climate, soils and geography.

Biofuel

Biofuels include a wide range of fuels which are derived from biomass. The term covers solid biomass, liquid fuels and various biogases. Liquid biofuels include bioalcohols, such as bioethanol, and oils, such as biodiesel. Gaseous biofuels include biogas, landfill gas and synthetic gas.

Bioethanol is an alcohol made by fermenting the sugar components of plant materials and it is made mostly from sugar and starch crops. With advanced technology being developed, cellulosic biomass, such as trees and grasses, are also used as feedstocks for ethanol production. Ethanol can be used as a fuel for vehicles in its pure form, but it is usually used as a gasoline additive to increase octane and improve vehicle emissions. Bioethanol is widely used in the USA and in Brazil.

Biodiesel is made from vegetable oils, animal fats or recycled greases. Biodiesel can be used as a fuel for vehicles in its pure form, but it is usually used as a diesel additive to reduce levels of particulates, carbon monoxide, and hydrocarbons from diesel-powered vehicles. Biodiesel is produced from oils or fats using transesterification and is the most common biofuel in Europe.

Biofuels provided 2.7% of the world's transport fuel in 2010.

Geothermal Energy

Geothermal energy is thermal energy generated and stored in the Earth. Thermal energy is the energy that determines the temperature of matter. Earth's geothermal energy originates from the original formation of the planet (20%) and from radioactive decay of minerals (80%). The geothermal gradient, which is the difference in temperature between the core of the planet and its surface, drives a continuous conduction of thermal energy in the form of heat from the core to the surface. The adjective *geothermal* originates from the Greek roots *geo*, meaning earth, and *thermos*, meaning heat.

The heat that is used for geothermal energy can be stored deep within the Earth, all the way down to Earth's core – 4,000 miles down. At the core, temperatures may reach over 9,000 degrees Fahrenheit. Heat conducts from the core to surrounding rock. Extremely high temperature and pressure cause some rock to melt, which is commonly known as magma. Magma convects upward since it is lighter than the solid rock. This magma then heats rock and water in the crust, sometimes up to 700 degrees Fahrenheit

From hot springs, geothermal energy has been used for bathing since Paleolithic times and for space heating since ancient Roman times, but it is now better known for electricity generation.

Renewable Energy Commercialisation

Growth of Renewables

During the five-years from the end of 2004 through 2009, worldwide renewable energy capacity grew at rates of 10–60% annually for many technologies. For wind power and many other renewable technologies, growth accelerated in 2009 relative to the previous four years. More wind power capacity was added during 2009 than any other renewable technology. However, grid-connected PV increased the fastest of all renewables technologies, with a 60% annual average growth rate for the five-year period. In 2010, renewable power consisted about a third of the newly built power generation capacities.

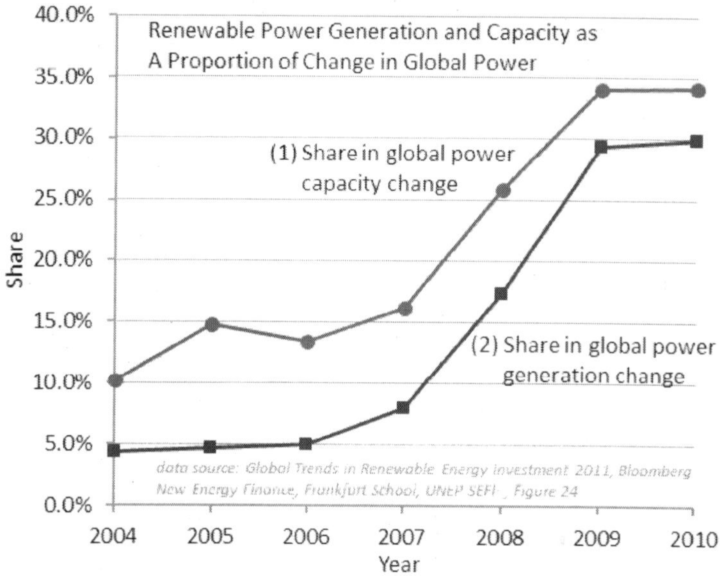

Figure : *Renewable power generation and capacity as a proportion of change in global power supply*

Scientists have advanced a plan to power 100% of the world's energy with wind, hydroelectric, and solar power by the year 2030.

According to a 2011 projection by the International Energy Agency, solar power generators may produce most of the world's electricity within 50 years, dramatically reducing the emissions of greenhouse gases that harm the environment. Cedric Philibert, senior analyst in the renewable energy division at the IEA said: "Photovoltaic and solar-thermal plants may meet most of the world's demand for electricity by 2060 — and half of all energy needs — with wind, hydropower and biomass plants supplying much of the remaining generation". "Photovoltaic and concentrated solar power together can become the major source of electricity," Philibert said.

Economic Trends

All forms of energy are expensive, but as time progresses, renewable energy generally gets cheaper, while fossil fuels generally get more expensive. A 2011 IEA report said: "A portfolio of renewable energy technologies is becoming cost-competitive in an increasingly broad range of circumstances, in some cases providing investment opportunities without the need for specific economic support," and added that "cost reductions in critical technologies, such as wind and solar, are set to continue."

The International Solar Energy Society argues that renewable energy technologies and economics will continue to improve with time, and that they are "sufficiently advanced at present to allow for major penetrations of renewable energy into the mainstream energy and societal infrastructures".

Hydroelectricity

The Three Gorges Dam in Hubei, China, has the world's largest instantaneous generating capacity (22,500 MW), with the Itaipu dam in Brazil/Paraguay in second place (14,000 MW). The giant Three Gorges Dam (22,500 MW when completed) is operated jointly with the much smaller Gezhouba Dam (3,115 MW). As of 2009, the total generating capacity of this two-dam complex is 21,515 MW. The whole project is planned to be completed in 2012, when the total generating capacity will be 25,615 MW. In 2008, this complex generated 97.9 TWh of electricity (80.8 TWh from the Three Gorges Dam and 17.1 TWh from the Gezhouba Dam), which is 3.4% more power in one year than the 94.7 TWh generated by Itaipu in 2008.

Wind Power Market

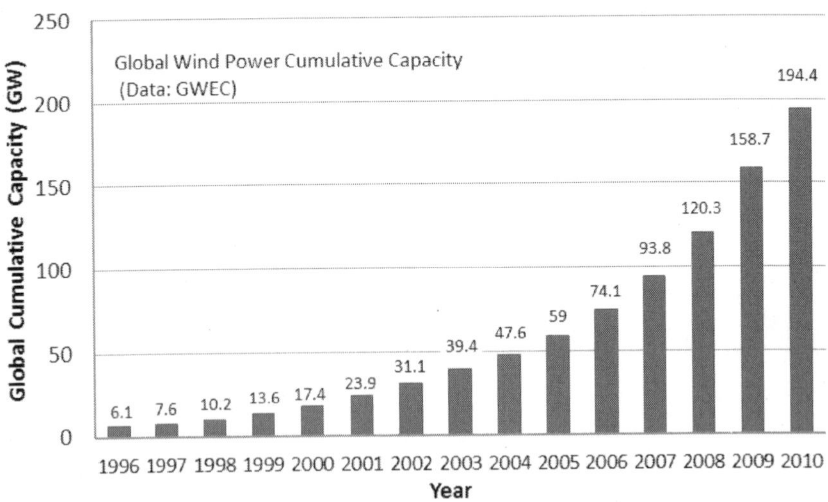

Figure: *Wind power: worldwide installed capacity*

Wind power is growing at the rate of 30% annually, with a worldwide installed capacity of 238,000 megawatts (MW) at the end of

2011, and is widely used in Europe, Asia, and the United States. Several countries have achieved relatively high levels of wind power penetration, such as 21% of stationary electricity production in Denmark, 18% in Portugal, 16% in Spain, 14% in Ireland and 9% in Germany in 2010. As of 2011, 83 countries around the world are using wind power on a commercial basis. As of 2011, the Roscoe Wind Farm (781 MW) is the world's largest wind farm. As of February 2012, the Walney Wind Farm in United Kingdom is the largest offshore wind farm in the world at 367 MW, followed by Thanet Offshore Wind Project (300 MW), also in the UK. The London Array (630 MW) is the largest project under construction. The United Kingdom is the world's leading generator of offshore wind power, followed by Denmark.

There are many large wind farms under construction and these include BARD Offshore 1 (400 MW), Clyde Wind Farm (548 MW), Greater Gabbard wind farm (500 MW), Lincs Wind Farm (270 MW), London Array (1000 MW), Lower Snake River Wind Project (343 MW), Macarthur Wind Farm (420 MW), Shepherds Flat Wind Farm (845 MW), and the Sheringham Shoal (317 MW).

New Generation of Solar Thermal Plants

Large solar thermal power stations include the 354 megawatt (MW) Solar Energy Generating Systems power plant in the USA, Solnova Solar Power Station (Spain, 150 MW), Andasol Solar Power Station (Spain, 100 MW), Nevada Solar One (USA, 64 MW), PS20 solar power plant (Spain, 20 MW), and the PS10 Solar Power Plant (Spain, 11 MW).

The Ivanpah Solar Power Facility is a 392 MW solar power facility which is under construction in south-eastern California. The Solana Generating Station is a 280 MW solar power plant which is under construction near Gila Bend, Arizona, about 70 miles (110 km) southwest of Phoenix. The Crescent Dunes Solar Energy Project is a 110 MW solar thermal power project currently under construction near Tonopah, about 190 miles (310 km) northwest of Las Vegas.

The solar thermal power industry is growing rapidly with 1.2 GW under construction as of April 2009 and another 13.9 GW announced globally through 2014. Spain is the epicentre of solar thermal power development with 22 projects for 1,037 MW under construction, all of which are projected to come online by the end of 2010. In the United States, 5,600 MW of solar thermal power projects have been announced. In developing countries, three World Bank projects for integrated solar thermal/combined-cycle gas-turbine power plants in Egypt, Mexico, and Morocco have been approved.

Photovoltaic Market

Solar photovoltaic cells convert sunlight into electricity and photovoltaic production has been increasing by an average of more than 20% each year since 2002, making it a fast-growing energy technology. At the end of 2011 the photovoltaic (PV) capacity worldwide was 67.4 GW. Top capacity countries were, in GW: Germany 24.7, Italy 12.5, Japan 4.7, Spain 4.2, the USA 4.2, and China 2.9.

Many solar photovoltaic power stations have been built, mainly in Europe. As of December 2011, the largest photovoltaic (PV) power plants in the world are the Golmud Solar Park (China, 200 MW), Sarnia Photovoltaic Power Plant (Canada, 97 MW), Montalto di Castro Photovoltaic Power Station (Italy, 84.2 MW), Finsterwalde Solar Park (Germany, 80.7 MW), Okhotnykovo Solar Park (Ukraine, 80 MW), Lieberose Photovoltaic Park (Germany, 71.8 MW), Rovigo Photovoltaic Power Plant (Italy, 70 MW), Olmedilla Photovoltaic Park (Spain, 60 MW), and the Strasskirchen Solar Park (Germany, 54 MW).

There are also many large plants under construction. The Desert Sunlight Project is a 550 MW solar power plant under construction in Riverside County, California, that will use thin-film solar photovoltaic modules made by First Solar. The Topaz Solar Farm is a 550 MW photovoltaic power plant, being built in San Luis Obispo County, California. The Blythe Solar Power Projectis a 500 MW photovoltaic station under construction in Riverside County, California. The Agua Caliente Solar Project is a 290 megawatt photovoltaic solar generating facility being built in Yuma County, Arizona. The California Valley Solar Ranch (CVSR) is a 250 megawatt (MW) solar photovoltaic power plant, which is being built by Sun Power in the Carrizo Plain, northeast of California Valley. The 230 MW Antelope Valley Solar Ranch is a First Solar photovoltaic project which is under construction in the Antelope Valley area of the Western Mojave Desert, and due to be completed in 2013.

Many of these plants are integrated with agriculture and some use innovative tracking systems that follow the sun's daily path across the sky to generate more electricity than conventional fixed-mounted systems. There are no fuel costs or emissions during operation of the power stations.

However, when it comes to renewable energy systems and PV, it is not just large systems that matter. Building-integrated photovoltaics or "onsite" PV systems use existing land and structures and generate power close to where it is consumed.

Biofuels for Transportation

Biofuels provided 3% of the world's transport fuel in 2010. Mandates for blending biofuels exist in 31 countries at the national level and in 29 states/provinces. According to the International Energy Agency, biofuels have the potential to meet more than a quarter of world demand for transportation fuels by 2050.

Since the 1970s, Brazil has had an ethanol fuel program which has allowed the country to become the world's second largest producer of ethanol (after the United States) and the world's largest exporter. Brazil's ethanol fuel program uses modern equipment and cheap sugarcane as feedstock, and the residual cane-waste (bagasse) is used to produce heat and power. There are no longer light vehicles in Brazil running on pure gasoline. By the end of 2008 there were 35,000 filling stations throughout Brazil with at least one ethanol pump.

Nearly all the gasoline sold in the United States today is mixed with 10% ethanol, a mix known as E10, and motor vehicle manufacturers already produce vehicles designed to run on much higher ethanol blends. Ford, Daimler AG, and GM are among the automobile companies that sell "flexible-fuel" cars, trucks, and minivans that can use gasoline and ethanol blends ranging from pure gasoline up to 85% ethanol (E85). By mid-2006, there were approximately 6 million E85-compatible vehicles on U.S. roads. The challenge is to expand the market for biofuels beyond the farm states where they have been most popular to date. Flex-fuel vehicles are assisting in this transition because they allow drivers to choose different fuels based on price and availability. The *Energy Policy Act of 2005*, which calls for 7.5 billion US gallons (28,000,000 m^3) of biofuels to be used annually by 2012, will also help to expand the market.

Geothermal Energy Commercialisation

Geothermal power is cost effective, reliable, sustainable, and environmentally friendly, but has historically been limited to areas near tectonic plate boundaries. Recent technological advances have dramatically expanded the range and size of viable resources, especially for applications such as home heating, opening a potential for widespread exploitation. Geothermal wells release greenhouse gases trapped deep within the earth, but these emissions are much lower per energy unit than those of fossil fuels. As a result, geothermal power has the potential to help mitigate global warming if widely deployed in place of fossil fuels.

The International Geothermal Association (IGA) has reported that 10,715 MW of geothermal power in 24 countries is online, which is expected to generate 67,246 GWh of electricity in 2010. This represents a 20% increase in geothermal power online capacity since 2005. IGA projects this will grow to 18,500 MW by 2015, due to the large number of projects presently under consideration, often in areas previously assumed to have little exploitable resource.

In 2010, the United States led the world in geothermal electricity production with 3,086 MW of installed capacity from 77 power plants; the largest group of geothermal power plants in the world is located at The Geysers, a geothermal field in California. The Philippines follows the US as the second highest producer of geothermal power in the world, with 1,904 MW of capacity online; geothermal power makes up approximately 18% of the country's electricity generation.

Developing Country Markets

Renewable energy can be particularly suitable for developing countries. In rural and remote areas, transmission and distribution of energy generated from fossil fuels can be difficult and expensive. Producing renewable energy locally can offer a viable alternative.

Technology advances are opening up a huge new market for solar power: the approximately 1.3 billion people around the world who don't have access to grid electricity. Even though they are typically very poor, these people have to pay far more for lighting than people in rich countries because they use inefficient kerosene lamps. Solar power costs half as much as lighting with kerosene. An estimated 3 million households get power from small solar PV systems. Kenya is the world leader in the number of solar power systems installed per capita. More than 30,000 very small solar panels, each producing 12 to 30 watts, are sold in Kenya annually.

Micro-hydro configured into mini-grids also provide power. Over 44 million households use biogas made in household-scale digesters for lighting and/or cooking, and more than 166 million households rely on a new generation of more-efficient biomass cookstoves.

Renewable energy projects in many developing countries have demonstrated that renewable energy can directly contribute to poverty alleviation by providing the energy needed for creating businesses and employment. Renewable energy technologies can also make indirect contributions to alleviating poverty by providing energy for cooking, space heating, and lighting. Renewable energy can also contribute to education, by providing electricity to schools.

Industry and Policy Trends

U.S. President Barack Obama's American Recovery and Reinvestment Act of 2009 includes more than $70 billion in direct spending and tax crs for clean energy and associated transportation programs. Clean Edge suggests that the commercialisation of clean energy will help countries around the world pull out of the current economic malaise.

Leading renewable energy companies include First Solar, Gamesa, GE Energy, Q-Cells, Sharp Solar, Siemens, SunOpta, Suntech, and Vestas. The military has also focused on the use of renewable fuels for military vehicles. Unlike fossil fuels, renewable fuels can be produced in any country, creating a strategic advantage. The US military has already committed itself to have 50% of its energy consumption come from alternative sources.

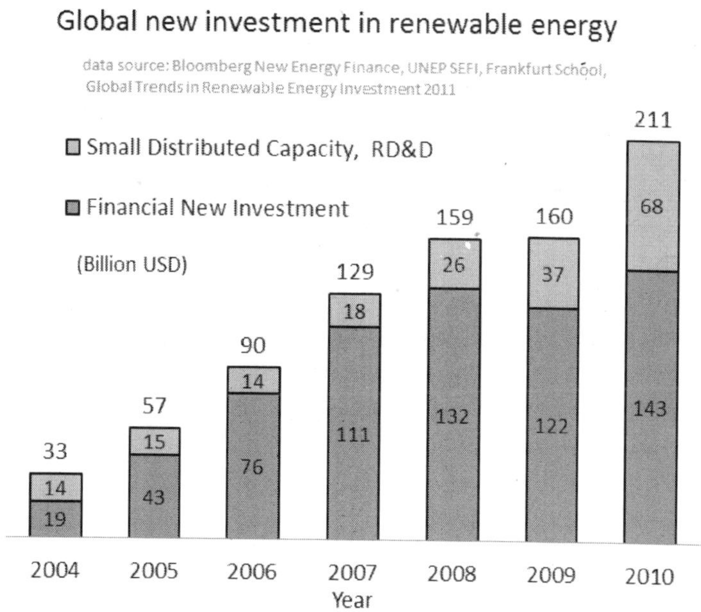

Figure : *Global New Investments in Renewable Energy*

The International Renewable Energy Agency (IRENA) is an intergovernmental organisation for promoting the adoption of renewable energy worldwide. It aims to provide concrete policy advice and facilitate capacity building and technology transfer. IRENA was formed on January 26, 2009, by 75 countries signing the charter of IRENA. As of March 2010, IRENA has 143 member states who all are considered as founding members, of which 14 have also ratified the statute. As of 2011, 119 countries have some form of national renewable energy policy target or renewable support policy. National targets now exist in at least 98 countries. There is also a wide range of policies at state/provincial and local levels. United Nations' Secretary-General Ban Ki-moon has said that renewable energy has the ability to lift the poorest nations to new levels of prosperity.

In October 2011, he "announced the creation of a high-level group to drum up support for energy access, energy efficiency and greater use of renewable energy. The group is to be co-chaired by Kandeh Yumkella, the chair of UN Energy and director general of the UN Industrial Development Organisation, and Charles Holliday, chairman of Bank of America".

Chapter 4

Petroleum Product

Petroleum products are useful materials derived from crude oil (petroleum) as it is processed in oil refineries. According to crude oil composition and demand, refineries can produce different shares of petroleum products. The largest share of oil products is used as energy carriers: various grades of fuel oil and gasoline. These energy-carrying fuels include or can be blended to give gasoline, jet fuel, diesel fuel, heating oil, and heavier fuel oils.

Figure: *A petrochemical refinery in Grangemouth, Scotland.*

Heavier (less volatile) fractions can also be used to produce asphalt, tar, paraffin wax, lubricating and other heavy oils. Refineries also produce other chemicals, some of which are used in chemical processes to produce plastics and other useful materials. Since petroleum often contains a couple of percent sulphur, sulphur is also often produced as a petroleum product. Hydrogen and carbon in the form of petroleum coke may also be produced as petroleum products. The hydrogen produced is often used as an intermediate product for other oil refinery processes such as hydrogen catalytic cracking and hydrodesulphurisation.

Specialty and Products

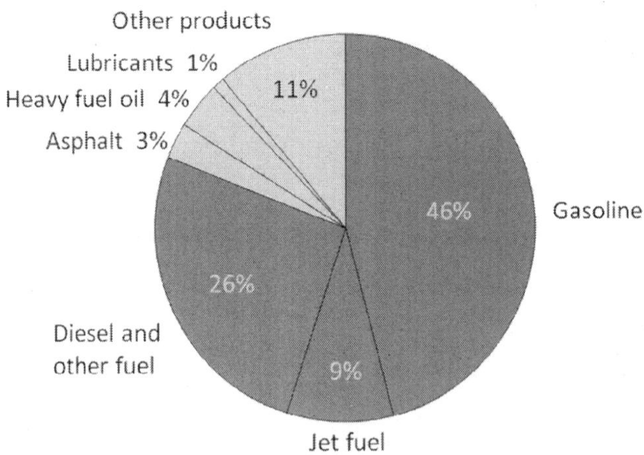

***Figure :** A breakdown of the products made from a typical barrel of US oil.*

Oil refineries will blend various feedstocks, mix appropriate additives, provide short term storage, and prepare for bulk loading to trucks, barges, product ships, and railcars.

- Gaseous fuels such as propane, stored and shipped in liquid form under pressure in specialised railcars to distributors.
- Liquid fuels blending (producing automotive and aviation grades of gasoline, kerosene, various aviation turbine fuels, and diesel fuels, adding dyes, detergents, antiknock additives, oxygenates, and anti-fungal compounds as required). Shipped by barge, rail, and tanker ship. May be shipped regionally in dedicated pipelines to point consumers, particularly aviation jet fuel to major airports, or piped to distributors in multi-product pipelines using product separators called pipeline inspection gauges ("pigs").

- Lubricants (produces light machine oils, motor oils, and greases, adding viscosity stabilisers as required), usually shipped in bulk to an offsite packaging plant.
- Wax (paraffin), used in the packaging of frozen foods, among others. May be shipped in bulk to a site to prepare as packaged blocks.
- Sulphur (or sulphuric acid), byproducts of sulphur removal from petroleum which may have up to a couple percent sulphur as organic sulphur-containing compounds. Sulphur and sulphuric acid are useful industrial materials. Sulphuric acid is usually prepared and shipped as the acid precursor oleum.
- Bulk tar shipping for offsite unit packaging for use in tar-and-gravel roofing or similar uses.
- Asphalt - used as a binder for gravel to form asphalt concrete, which is used for paving roads, lots, etc. An asphalt unit prepares bulk asphalt for shipment.
- Petroleum coke, used in speciality carbon products such as certain types of electrodes, or as solid fuel.
- Petrochemicals or petrochemical feedstocks, which are often sent to petrochemical plants for further processing in a variety of ways. The petrochemicals may be olefins or their precursors, or various types of aromatic petrochemicals.

Petrochemicals have a vast variety of uses. They are commonly used as monomers or feedstocks for monomer production. Olefins such as alpha-olefins anddienes are often used as monomers, although aromatics can also be used as monomer precursors.

The monomers are then polymerised in various ways to form polymers. Polymer materials can be used as plastics, elastomers, or fibres, or possibly some intermediate form of these material types. Some polymers are also used as gels or lubricants. Petrochemicals can also be used as solvents or as feedstock for producing solvents. Petrochemicals can also be used as precursors for a wide variety of chemicals and substances such as vehicle fluids, surfactants for cleaners, etc.

Etymology

The word *petroleum* comes from Greek "petra" for rock and "elaion" for oil. The term was found (in the spelling "petraoleum") in 10th-century Old English sources. It was used in the treatise *De Natura Fossilium*, published in 1546 by the German mineralogist Georg Bauer, also known as Georgius Agricola. In the 19th century, the term*petroleum* was frequently used to refer to mineral oils produced by distillation from mined organic solids such as cannel coal (and later oil shale), and refined oils produced from them; in the United Kingdom, storage (and later transport) of these oils were regulated by a series of Petroleum Acts, from the *Petroleum Act 1862* c. 66 onward.

Composition

In its strictest sense, petroleum includes only crude oil, but in common usage it includes all liquid, gaseous, and solid (e.g., paraffin) hydrocarbons. Under surface pressure and temperature conditions, lighter hydrocarbons methane, ethane, propane and butane occur as gases, while pentane and heavier ones are in the form of liquids or solids. However, in an underground oil reservoir the proportions of gas, liquid, and solid depend on subsurface conditions and on the phase diagram of the petroleum mixture.

An oil well produces predominantly crude oil, with some natural gas dissolved in it. Because the pressure is lower at the surface than underground, some of the gas will come out of solution and be recovered (or burned) as *associated gas* or *solution gas*. A gas well produces predominantly natural gas. However, because the underground temperature and pressure are higher than at the surface, the gas may contain heavier hydrocarbons such as pentane, hexane, and heptane in the gaseous state. At surface conditions these will condense out of the gas to form natural gas condensate, often shortened to *condensate.* Condensate resembles petrol in appearance and is similar in composition to some volatile light crude oils.

The proportion of light hydrocarbons in the petroleum mixture varies greatly among different oil fields, ranging from as much as 97% by weight in the lighter oils to as little as 50% in the heavier oils and bitumens. The hydrocarbons in crude oil are mostly alkanes, cycloalkanes and various aromatic hydrocarbons while the other organic compounds contain nitrogen, oxygen and sulphur, and trace amounts of metals such as iron, nickel, copper and vanadium. The exact molecular composition varies widely from formation to formation but the proportion of chemical elements vary over fairly narrow limits as follows:

Four different types of hydrocarbon molecules appear in crude oil. The relative percentage of each varies from oil to oil, determining the properties of each oil. Crude oil varies greatly in appearance depending on its composition. It is usually black or dark brown (although it may be yellowish, reddish, or even greenish). In the reservoir it is usually found in association with natural gas, which being lighter forms a gas cap over the petroleum, and saline water which, being heavier than most forms of crude oil, generally sinks beneath it. Crude oil may also be found in semi-solid form mixed with sand and water, as in the Athabasca oil sands in Canada, where it is usually referred to as crude bitumen. In Canada, bitumen is considered a sticky, black, tar-like form of crude oil which is so thick and heavy that it must be heated or diluted before it will flow. Venezuela also has large amounts of oil in the Orinoco oil sands, although the hydrocarbons trapped in them are more fluid than in Canada and are usually called extra heavy oil. These oil sands resources are called unconventional oil to distinguish them from oil which can be extracted using traditional oil well methods. Between them, Canada and Venezuela contain an estimated 3.6 trillion barrels (570×10^9 m^3) of bitumen and extra-heavy oil, about twice the volume of the world's reserves of conventional oil.

Petroleum is used mostly, by volume, for producing fuel oil and petrol, both important *"primary energy"* sources. 84% by volume of the hydrocarbons present in petroleum is converted into energy-rich fuels (petroleum-based fuels), including petrol, diesel, jet, heating, and other fuel oils, and liquefied petroleum gas. The lighter grades of crude oil produce the best yields of these products, but as the world's reserves of light and medium oil are depleted, oil refineries are increasingly having to process heavy oil and bitumen, and use more complex and expensive methods to produce the products required. Because heavier crude oils have too much carbon and not enough hydrogen, these processes generally involve removing carbon from or adding hydrogen to the molecules, and using fluid catalytic cracking to convert the longer, more complex molecules in the oil to the shorter, simpler ones in the fuels.

Due to its high energy density, easy transportability and relative abundance, oil has become the world's most important source of energy since the mid-1950s. Petroleum is also the raw material for many chemical products, including pharmaceuticals, solvents, fertilizers, pesticides, and plastics; the 16% not used for energy production is converted into these other materials. Petroleum is found in porous rock formations in the upper strata of some areas of the Earth's crust. There is also petroleum in oil sands (tar sands). Known oil reserves are typically estimated at around 190 km^3 (1.2 trillion (short scale) barrels) without oil sands, or 595 km^3 (3.74 trillion barrels) with oil sands. Consumption is currently around 84 million barrels (13.4×10^6 m^3) per day, or 4.9 km^3 per year. Which in turn yields a remaining oil supply of only about 120 years, if current demand remain static.

Chemistry

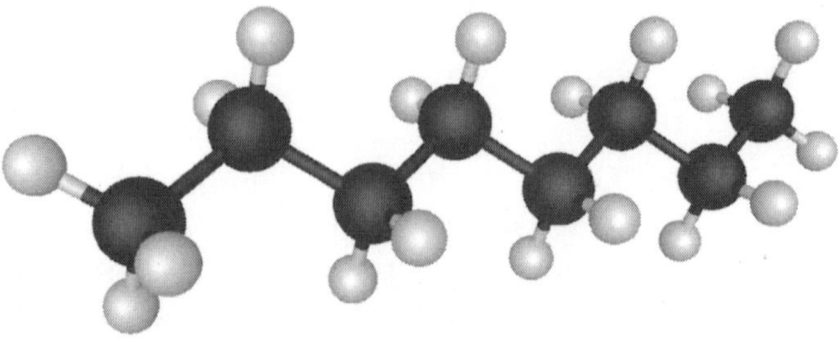

Figure : *Octane, a hydrocarbon found in petroleum. Lines represent single bonds; black spheres represent carbon; white spheres represent hydrogen.*

Petroleum is a mixture of a very large number of different hydrocarbons; the most commonly found molecules are alkanes (linear or branched), cycloalkanes, aromatic hydrocarbons, or more complicated chemicals like asphaltenes. Each petroleum variety has a unique mix of molecules, which define its physical and chemical properties, like colour and viscosity.

The *alkanes*, also known as *paraffins*, are saturated hydrocarbons with straight or branched chains which contain only carbon and hydrogen and have the general formula C_nH_{2n+2}. They generally have from 5 to 40 carbon atoms per molecule, although trace amounts of shorter or longer molecules may be present in the mixture.

The alkanes from pentane (C_5H_{12}) to octane (C_8H_{18}) are refined into petrol, the ones from nonane (C_9H_{20}) to hexadecane ($C_{16}H_{34}$) into diesel fuel, kerosene and jet fuel. Alkanes with more than 16 carbon atoms can be refined into fuel oil and lubricating oil. At the heavier end of the range, paraffin wax is an alkane with approximately 25 carbon atoms, while asphalt has 35 and up, although these are usually cracked by modern refineries into more valuable products. The shortest molecules, those with four or fewer carbon atoms, are in a gaseous state at room temperature. They are the petroleum gases. Depending on demand and the cost of recovery, these gases are either flared off, sold as liquified petroleum gas under pressure, or used to power the refinery's own burners. During the winter, Butane (C_4H_{10}), is blended into the petrol pool at high rates, because butane's high vapour pressure assists with cold starts. Liquified under pressure slightly above atmospheric, it is best known for powering cigarette lighters, but it is also a main fuel source for many developing countries. Propane can be liquified under modest pressure, and is consumed for just about every application relying on petroleum for energy, from cooking to heating to transportation.

The *cycloalkanes*, also known as *naphthenes*, are saturated hydrocarbons which have one or more carbon rings to which hydrogen atoms are attached according to the formula C_nH_{2n}. Cycloalkanes have similar properties to alkanes but have higher boiling points.

The *aromatic hydrocarbons* are unsaturated hydrocarbons which have one or more planar six-carbon rings called benzene rings, to which hydrogen atoms are attached with the formula C_nH_n. They tend to burn with a sooty flame, and many have a sweet aroma. Some are carcinogenic.

These different molecules are separated by fractional distillation at an oil refinery to produce petrol, jet fuel, kerosene, and other hydrocarbons. For example, 2,2,4-Trimethylpentane (isooctane), widely used in petrol, has a chemical formula of C_8H_{18} and it reacts with oxygen exothermically:

$$2\ C_8H_{18(l)} + 25\ O_{2(g)} \rightarrow 16\ CO_{2(g)} + 18\ H_2O_{(g)} + 10.86\ \text{MJ/mol (of octane)}$$

The amount of various molecules in an oil sample can be determined in laboratory. The molecules are typically extracted in a solvent, then separated in a gas chromatograph, and finally determined with a suitable detector, such as a flame ionisation detector or a mass spectrometre. Due to the large number of co-eluted hydrocarbons within oil, many cannot be resolved by traditional gas chromatography and typically appear as a hump in the chromatogram. This unresolved complex mixture (UCM) of hydrocarbons is particularly apparent when analysing weathered oils and extracts from tissues of organisms exposed to oil. Incomplete combustion of petroleum or petrol results in production of toxic byproducts. Too little oxygen results in carbon monoxide. Due to the high temperatures and high pressures involved, exhaust gases from petrol combustion in car engines usually include nitrogen oxides which are responsible for creation of photochemical smog.

Empirical Equations for Thermal Properties

Heat of Combustion

At a constant volume the heat of combustion of a petroleum product can be approximated as follows:

$$Q_v = 12{,}400 - 2{,}100d^2.$$

where Q_v is measured in cal/gram and d is the specific gravity at 60 °F (16 °C).

Thermal Conductivity

The thermal conductivity of petroleum based liquids can be modelled as follows:

$$K = \frac{0.813}{d}[1 - 0.0203(t - 32)] \qquad 0.547$$

where K is measured in BTU $\cdot hr^{-1}ft^{-2}$, t is measured in °F and d is the specific gravity at 60 °F (16 °C).

Specific Heat

The specific heat of a petroleum oils can be modelled as follows:

$$c = \frac{1}{\sqrt{d}}[0.388 + 0.00045t],$$

where c is measured in BTU/lbm-°F, t is the temperature in Fahrenheit and *d* is the specific gravity at 60 °F (16 °C).

In units of kcal/(kg·°C), the formula is:

$$\frac{1}{\sqrt{d}}[0.402 + 0.00081t]$$

where the temperature *t* is in Celsius and *d* is the specific gravity at 15 °C.

Latent Heat of Vapourisation

The latent heat of vapourisation can be modelled under atmospheric conditions as follows:

$$L = \frac{1}{d}[110.9 - 0.09t]$$

where L is measured in BTU/lbm, t is measured in °F and d is the specific gravity at 60 °F (16 °C).

In units of kcal/kg, the formula is:

$$L = \frac{1}{d}[194.4 - 0.162t]$$

where the temperature *t* is in Celsius and *d* is the specific gravity at 15 °C.

Formation

Petroleum is a fossil fuel derived from ancient fossilised organic materials, such as zooplankton and algae. Vast quantities of these remains settled to sea or lake bottoms, mixing with sediments and being buried under anoxic conditions. As further layers settled to the sea or lake bed, intense heat and pressure built up in the lower regions. This process caused the organic matter to change, first into a waxy material known as kerogen, which is found in various oil shales around the world, and then with more heat into liquid and gaseous hydrocarbons via a process known as catagenesis. Formation of petroleum occurs from hydrocarbon pyrolysis in a variety of mainly endothermic reactions at high temperature and/or pressure.

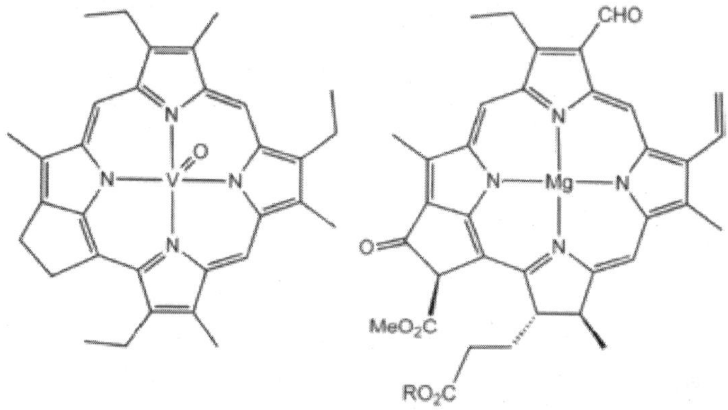

Figure : *Structure of vanadium porphyrin compound extracted from petroleum by Alfred E. Treibs, father of organic geochemistry. Treibs noted the close structural similarity of this molecule and chlorophyll a.*

There were certain warm nutrient-rich environments such as the Gulf of Mexico and the ancient Tethys Sea where the large amounts of organic material falling to the ocean floor exceeded the rate at which it could decompose. This resulted in large masses of organic material being buried under subsequent deposits such as shale formed from mud. This massive organic deposit later became heated and transformed under pressure into oil.

Geologists often refer to the temperature range in which oil forms as an “oil window”—below the minimum temperature oil remains trapped in the form of kerogen, and above the maximum temperature the oil is converted to natural gas through the process of thermal cracking. Sometimes, oil formed at extreme depths may migrate and become trapped at a much shallower level. The Athabasca Oil Sands is one example of this.

Reservoirs

Crude Oil Reservoirs

Three conditions must be present for oil reservoirs to form: a source rock rich in hydrocarbon material buried deep enough for subterranean heat to cook it into oil; a porous and permeable reservoir rock for it to accumulate in; and a cap rock (seal) or other mechanism that prevents it from escaping to the surface. Within these reservoirs, fluids will typically organise themselves like a three-layer cake with a layer of water below the oil layer and a layer of gas above it, although the different layers vary in size between reservoirs. Because most hydrocarbons are less dense than rock or water, they often migrate upward through adjacent rock layers until either reaching the surface or becoming trapped within porous rocks (known as reservoirs) by impermeable rocks above. However, the process is influenced by underground water flows, causing oil to migrate hundreds of kilometres horizontally or even short distances downward before becoming trapped in a reservoir. When hydrocarbons are concentrated in a trap, an oil field forms, from which the liquid can be extracted by drilling and pumping.

The reactions that produce oil and natural gas are often modelled as first order breakdown reactions, where hydrocarbons are broken down to oil and natural gas by a set of parallel reactions, and oil eventually breaks down to natural gas by another set of reactions. The latter set is regularly used in petrochemical plants and oil refineries.

Wells are drilled into oil reservoirs to extract the crude oil. "Natural lift" production methods that rely on the natural reservoir pressure to force the oil to the surface are usually sufficient for a while after reservoirs are first tapped. In some reservoirs, such as in the Middle East, the natural pressure is sufficient over a long time. The natural pressure in many reservoirs, however, eventually dissipates. Then the oil must be pumped out using "artificial lift" created by mechanical pumps powered by gas or electricity. Over time, these "primary" methods become less effective and "secondary" production methods may be used. A common secondary method is "waterflood" or injection of water into the reservoir to increase pressure and force the oil to the drilled shaft or "wellbore." Eventually "tertiary" or "enhanced" oil recovery methods may be used to increase the oil's flow characteristics by injecting steam, carbon dioxide and other gases or chemicals into the reservoir. In the United States, primary production methods account for less than 40% of the oil produced on a daily basis, secondary methods account for about half, and tertiary recovery the remaining 10%. Extracting oil (or "bitumen") from oil/tar sand and oil shale deposits requires mining the sand or shale and heating it in a vessel or retort, or using "in-situ" methods of injecting heated liquids into the deposit and then pumping out the oil-saturated liquid.

Unconventional Oil Reservoirs

Oil-eating bacteria biodegrade oil that has escaped to the surface. Oil sands are reservoirs of partially biodegraded oil still in the process of escaping and being biodegraded, but they contain so much migrating oil that, although most of it has escaped, vast amounts are still present—more than can be found in conventional oil reservoirs. The lighter fractions of the crude oil are destroyed first, resulting in reservoirs containing an extremely heavy form of crude oil, called crude bitumen in Canada, or extra-heavy crude oil in Venezuela. These two countries have the world's largest deposits of oil sands.

On the other hand, oil shales are source rocks that have not been exposed to heat or pressure long enough to convert their trapped hydrocarbons into crude oil. Technically speaking, oil shales are not always shales and do not contain oil, but are fined-grain sedimentary rocks containing an insoluble organic solid called kerogen. The kerogen in the rock can be converted into crude oil using heat and pressure to simulate natural processes. The method has been known for centuries and was patented in 1694 under British Crown Patent No. 330 covering, "A way to extract and make great quantities of pitch, tar, and oil out of a sort of stone." Although oil shales are found in many countries, the United States has the world's largest deposits.

Classification

Figure : *A sample of medium heavy crude oil*

The petroleum industry generally classifies crude oil by the geographic location it is produced in (e.g. West Texas Intermediate, Brent, or Oman), its API gravity (an oil industry measure of density), and its sulphur content. Crude oil may be considered *light* if it has low density or *heavy* if it has high density; and it may be referred to as sweet if it contains relatively little sulphur or *sour* if it contains substantial amounts of sulphur.

The geographic location is important because it affects transportation costs to the refinery. *Light* crude oil is more desirable than *heavy* oil since it produces a higher yield of petrol, while *sweet* oil commands a higher price than *sour* oil because it has fewer environmental problems and requires less refining to meet sulphur standards imposed on fuels in consuming countries. Each crude oil has unique molecular characteristics which are understood by the use of crude oil assay analysis in petroleum laboratories.

Barrels from an area in which the crude oil's molecular characteristics have been determined and the oil has been classified are used as pricing references throughout the world. Some of the common reference crudes are:

- West Texas Intermediate (WTI), a very high-quality, sweet, light oil delivered at Cushing, Oklahoma for North American oil
- Brent Blend, comprising 15 oils from fields in the Brent and Ninian systems in the East Shetland Basin of the North Sea. The oil is landed at Sullom Voe terminal in Shetland. Oil production from Europe, Africa and Middle Eastern oil flowing West tends to be priced off this oil, which forms a benchmark
- Dubai-Oman, used as benchmark for Middle East sour crude oil flowing to the Asia-Pacific region
- Tapis (from Malaysia, used as a reference for light Far East oil)
- Minas (from Indonesia, used as a reference for heavy Far East oil)
- The OPEC Reference Basket, a weighted average of oil blends from various OPEC (The Organisation of the Petroleum Exporting Countries) countries
- Midway Sunset Heavy, by which heavy oil in California is priced

There are declining amounts of these benchmark oils being produced each year, so other oils are more commonly what is actually delivered. While the reference price may be for West Texas Intermediate delivered at Cushing, the actual oil being traded may be a discounted Canadian heavy oil delivered at Hardisty, Alberta, and for a Brent Blend delivered at Shetland, it may be a Russian Export Blend delivered at the port of Primorsk.

Petroleum Industry

The petroleum industry is involved in the global processes of exploration, extraction, refining, transporting (often with oil tankers and pipelines), and marketing petroleum products. The largest volume products of the industry are fuel oil and petrol. Petroleum is also the raw material for many chemical products, including pharmaceuticals, solvents, fertilizers, pesticides, and plastics.

The industry is usually divided into three major components: upstream, midstream and downstream. Midstream operations are usually included in the downstream category. Petroleum is vital to many industries, and is of importance to the maintenance of industrialised civilization itself, and thus is critical concern to many nations. Oil accounts for a large percentage of the world's energy consumption, ranging from a low of 32% for Europe and Asia, up to a high of 53% for the Middle East. Other geographic regions' consumption patterns are as follows: South and Central America (44%), Africa (41%), and North America (40%). The world at large consumes 30 billion barrels (4.8 km^3) of oil per year, and the top oil consumers largely consist of developed nations. In fact, 24% of the oil consumed in 2004 went to the United States alone, though by 2007 this had dropped to 21% of world oil consumed.

In the US, in the states of Arizona, California, Hawaii, Nevada, Oregon and Washington, the Western States Petroleum Association (WSPA) represents companies responsible for producing, distributing, refining, transporting and marketing petroleum. This non-profit trade association was founded in 1907, and is the oldest petroleum trade association in the United States.

History

Petroleum, in one form or another, has been used since ancient times, and is now important across society, including in economy, politics and technology. The rise in importance was mostly due to the invention of the internal combustion engine, the rise in commercial aviation and the increasing use of plastic.

More than 4000 years ago, according to Herodotus and Diodorus Siculus, asphalt was used in the construction of the walls and towers of Babylon; there were oil pits near Ardericca (near Babylon), and a pitch spring on Zacynthus. Great quantities of it were found on the banks of the river Issus, one of the tributaries of the Euphrates. Ancient Persian tablets indicate the medicinal and lighting uses of petroleum in the upper levels of their society. By 347 AD, oil was produced from bamboo-drilled wells in China.

In the 1850s, the process to distill kerosene from petroleum was invented by Ignacy £ukasiewicz, providing a cheaper alternative to whale oil. The demand for the petroleum as a fuel for lighting in North America and around the world quickly grew. The world's first commercial oil well was drilled in Poland in 1853. Oil exploration developed in many parts of the world with the Russian Empire, particularly the Branobel company in Azerbaijan (Asia's first modern borehole oil production began in 1848 at the Bibi-Heybat field near Baku), taking the lead in production by the end of the 19th century.

Figure : *Oil derrick in Okemah, Oklahoma, 1922.*

Access to oil was and still is a major factor in several military conflicts of the twentieth century, including World War II, during which oil facilities were a major strategic asset and were extensively bombed. Operation Barbarossa included the goal to capture the Baku oilfields, as it would provide much needed oil-supplies for the German military which was suffering from blockades. Oil exploration in North America during the early 20th century later led to the U.S. becoming the leading producer by the mid 1900s. As petroleum production in the U.S. peaked during the 1960s, however, the United States was surpassed by Saudi Arabia and Russia.

Today, about 90% of vehicular fuel needs are met by oil. Petroleum also makes up 40% of total energy consumption in the United States, but is responsible for only 1% of electricity generation. Petroleum's worth as a portable, dense energy source powering the vast majority of vehicles and as the base of many industrial chemicals makes it one of the world's most important commodities. Viability of the oil commodity is controlled by several key parameters, number of vehicles in the world competing for fuel, quantity of oil exported to the world market (Export Land Model), Net Energy Gain (economically useful energy provided minus energy consumed), political stability of oil exporting nations and ability to defend oil supply lines.

The top three oil producing countries are Saudi Arabia, Russia, and the United States. About 80% of the world's readily accessible reserves are located in the Middle East, with 62.5% coming from the Arab 5: Saudi Arabia, UAE, Iraq, Qatar and Kuwait. A large portion of the world's total oil exists as unconventional sources, such as bitumen in Canada and Venezuela and oil shale. While significant volumes of oil are extracted from oil sands, particularly in Canada, logistical and technical hurdles remain, as oil extraction requires large amounts of heat and water, making its net energy content quite low relative to conventional crude oil. Thus, Canada's oil sands are not expected to provide more than a few million barrels per day in the foreseeable future. Conventional crude oil production, those having Net Energy Gain above 10 stopped growing in 2005 at about 74 million barrels per day (11,800,000 m^3/d). The International Energy Agency's (IEA) 2010 World Energy Outlook estimated that conventional crude oil production has peaked and is depleting at 6.8% per year. US Joint Forces Command's Joint Operating Environment 2010 issued this warning to all US military commands "By 2012, surplus oil production capacity could entirely disappear, and as early as 2015, the shortfall in output could reach nearly 10 million barrels per day."

Price

After the collapse of the OPEC-administered pricing system in 1985, and a short lived experiment with netback pricing, oil-exporting countries adopted a market-linked pricing mechanism. First adopted by PEMEX in 1986, market-linked pricing was widely accepted, and by 1988 became and still is the main method for pricing crude oil in international trade. The current reference, or pricing markers, are Brent, WTI, and Dubai/Oman.

Uses

The chemical structure of petroleum is heterogeneous, composed of hydrocarbon chains of different lengths. Because of this, petroleum may be taken to oil refineries and the hydrocarbon chemicals separated by distillation and treated by other chemical processes, to be used for a variety of purposes.

Fuels

A poster used to promote carpoolingas a way to ration gasoline during World War II. The most common distillation fractions of petroleum are fuels. Fuels include (by increasing boiling temperature range):

Table: Common Fractions of Petroleum as Fuels

Fraction	*Boiling Range °C*
Liquefied petroleum gas (LPG)	“40
Butane	“12 to “1
Petrol	“1 to 180
Jet fuel	150 to 205
Kerosene	205 to 260
Fuel oil	205 to 290
Diesel fuel	260 to 315

Other Derivatives

Certain types of resultant hydrocarbons may be mixed with other non-hydrocarbons, to create other end products:

- Alkenes (olefins) which can be manufactured into plastics or other compounds
- Lubricants (produces light machine oils, motor oils, and greases, adding viscosity stabilisers as required).
- Wax, used in the packaging of frozen foods, among others.
- Sulphur or Sulphuric acid. These are a useful industrial materials. Sulphuric acid is usually prepared as the acid precursor oleum, a byproduct of sulphur removal from fuels.
- Bulk tar.
- Asphalt
- Petroleum coke, used in speciality carbon products or as solid fuel.
- Paraffin wax
- Aromatic petrochemicals to be used as precursors in other chemical production.

Environmental Effects

Because petroleum is a naturally occurring substance, its presence in the environment need not be the result of human causes such as accidents and routine activities (seismic exploration, drilling, extraction, refining and combustion). Phenomena such as seeps and tar pits are examples of areas that petroleum naturally effects. Regardless of source, petroleum's effects when released into the environment are similar.

***Figure :** Diesel fuel spill on a road*

Global Warming

When burned, petroleum releases carbon dioxide; a greenhouse gas. Along with the burning of coal, petroleum combustion is the largest contributor to the increase in atmospheric CO_2. Atmospheric CO_2 has risen steadily since the industrial revolution to current levels of over 380ppmv, from the 180 – 300ppmv of the prior 800 thousand years, driving global warming.

Extraction

Oil extraction is simply the removal of oil from the reservoir (oil pool). Oil is often recovered as a water-in-oil emulsion, and speciality chemicals called demulsifiers are used to separate the oil from water. Oil extraction is costly and sometimes environmentally damaging, although Dr. John Hunt of the Woods Hole Oceanographic Institution pointed out in a 1981 paper that over 70% of the reserves in the world are associated with visible macroseepages, and many oil fields are found due to natural seeps. Offshore exploration and extraction of oil disturbs the surrounding marine environment.

Oil Spills

Crude oil and refined fuel spills from tanker ship accidents have damaged natural ecosystems in Alaska, the Gulf of Mexico, the Galapagos Islands, France and many other places.

The quantity of oil spilled during accidents has ranged from a few hundred tons to several hundred thousand tons (e.g., Deepwater Horizon Oil Spill, Atlantic Empress, Amoco Cadiz). Smaller spills have already proven to have a great impact on ecosystems, such as the Exxon Valdez oil spill

Oil spills at sea are generally much more damaging than those on land, since they can spread for hundreds of nautical miles in a thin oil slick which can cover beaches with a thin coating of oil. This can kill sea birds, mammals, shellfish and other organisms it coats. Oil spills on land are more readily containable if a makeshift earth dam can be rapidly bulldozed around the spill site before most of the oil escapes, and land animals can avoid the oil more easily. Control of oil spills is difficult, requires ad hoc methods, and often a large amount of manpower. The dropping of bombs and incendiary devices from aircraft on the Torrey Canyon wreck produced poor results; modern techniques would include pumping the oil from the wreck, like in the Prestige oil spill or the Erika oil spill. Though crude oil is predominantly composed of various hydrocarbons, certain nitrogen heterocylic compounds, such as pyridine, picoline, and quinoline are reported as contaminants associated with crude oil, as well as facilities processing oil shale or coal, and have also been found at legacy wood treatment sites. These compounds have a very high water solubility, and thus tend to dissolve and move with water. Certain naturally occurring bacteria, such as Micrococcus, Arthrobacter, and Rhodococcus and have been shown to degrade these contaminants.

Degradation of underground tanks located at gas stations may also occur, significantly affecting soil and groundwater. In these instances of the biological degradation of petroleum hydrocarbons, solutions such as bioremediation may be used.

Tarballs

A tarball is a blob of oil (not to be confused with tar, which is typically derived from pine trees rather than petroleum) which has been weathered after floating in the ocean. Tarballs are an aquatic pollutant in most environments, although they can occur naturally, for example, in the Santa Barbara Channel of California. Their concentration and features have been used to assess the extent of oil spills. Their composition can be used to identify their sources of origin, and tarballs themselves may be dispersed over long distances by deep sea currents. They are slowly decomposed by bacteria, including Chromobacterium violaceum, Cladosporium resinae, Bacillus submarinus, Micrococcus varians, Pseudomonas aeruginosa, Candida marina and Saccharomyces estuari.

Whales

James S. Robbins has argued that the advent of petroleum-refined kerosene saved some species of great whales from extinction by providing an inexpensive substitute for whale oil, thus eliminating the economic imperative for open-boat whaling.

Alternatives to Petroleum

Renewable Energy: In the United States in 2007 about 70% of petroleum was used for transportation (e.g. petrol, diesel, jet fuel), 24% by industry (e.g. production of plastics), 5% for residential and commercial uses, and 2% for electricity production. Outside of the US, a higher proportion of petroleum tends to be used for electricity.

Alternatives to Petroleum-based Vehicle Fuels

Alternative fuel vehicles refers to both:

- vehicles that use alternative fuels used in standard or modified internal combustion engines such as natural gas vehicles, neat ethanol vehicles, flexible-fuel vehicles, biodiesel-powered vehicles, and hydrogen vehicles.
- vehicles with advanced propulsion systems that reduce or substitute petroleum use such as battery electric vehicles, plug-in hybrid electric vehicles, hybrid electric vehicles, and hydrogen fuel cell vehicles.

Alternatives to Burning Petroleum for Electricity

In oil producing countries with little refinery capacity, oil is sometimes burned to produce electricity. Renewable energy technologies such as solar power, wind power, micro hydro, biomass and biofuels might someday be used to replace some of these generators, but today the primary alternatives remain large scale hydroelectricity, nuclear and coal-fired generation.

Future of Petroleum Production

Consumption in the twentieth and twenty-first centuries has been abundantly pushed by automobile growth; the 1985–2003 oil glut even fuelled the sales of low economy vehicles in OECD countries. In 2008, the economic crisis seems to have some impact on the sales of such vehicles; still, the 2008 oil consumption shows a small increase. The BRIC countries might also kick in, as China briefly was the first automobile market in December 2009.

The immediate outlook still hints upwards. In the long term, uncertainties linger; the OPEC believes that the OECD countries will push low consumption policies at some point in the future; when that happens, it will definitely curb the oil sales, and both OPEC and EIA kept lowering their 2020 consumption estimates during the past 5 years.

Oil products are more and more in competition with alternative sources, mainly coal and natural gas, both cheaper sources.

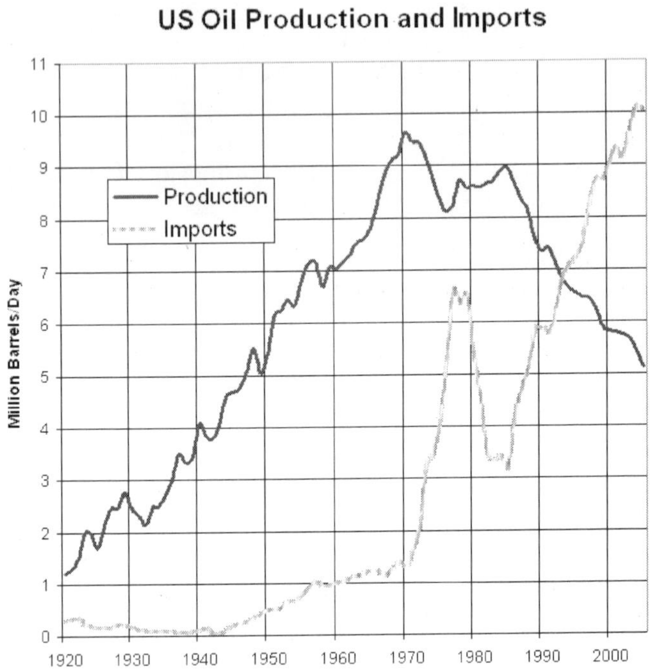

Figure : *US oil production and imports, 1920–2005.*

Production will also face an increasingly complex situation; while OPEC countries still have large reserves at low production prices, newly found reservoirs often lead to higher prices; offshore giants such as Tupi, Guara and Tiber demand high investments and ever-increasing technological abilities. Subsalt reservoirs such as Tupi were unknown in the twentieth century, mainly because the industry was unable to probe them. Enhanced Oil Recovery (EOR) techniques (example: DaQing, China) will continue to play a major role in increasing the world's recoverable oil.

Peak Oil

Peak Oil is the scientific projection that future petroleum production (whether for individual oil wells, entire oil fields, whole countries, or worldwide production) will eventually peak and then decline at a similar rate to the rate of increase before the peak as these reserves are exhausted. The peak of oil discoveries was in 1965, and oil production per year has surpassed oil discoveries every year since 1980.

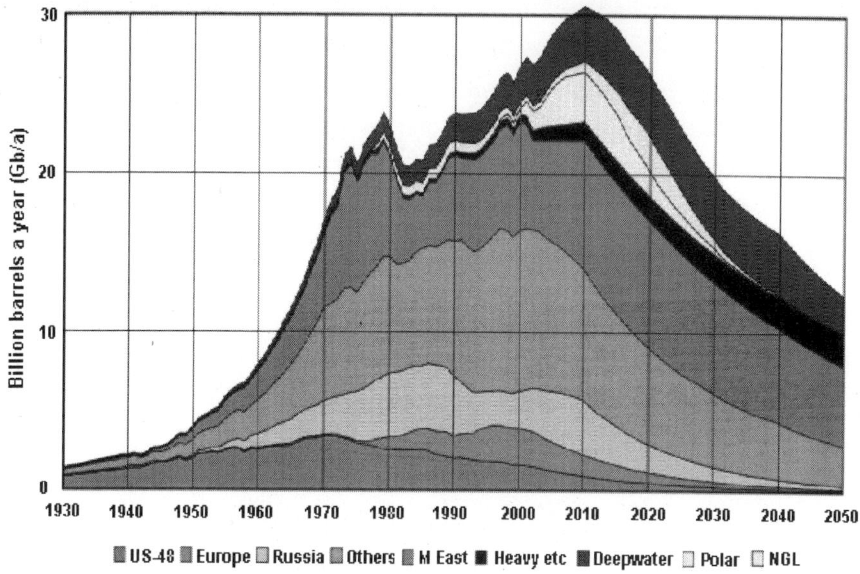

Figure : *Global Peak Oil forecast*

Hubbert applied his theory to accurately predict the peak of U.S. oil production at a date between 1966 and 1970. This prediction was based on data available at the time of his publication in 1956. In the same paper, Hubbert predicts world peak oil in "half a century" after his publication, which would be 2006.

It is difficult to predict the oil peak in any given region, due to the lack of knowledge and/or transparency in accounting of global oil reserves. The scientist and researchers from Oxford University argue that official figures are inflated because OPEC members over-reported reserves in the 1980s when competing for global market share. Based on available production data, proponents have previously predicted the peak for the world to be in years 1989, 1995, or 1995–2000. Some of these predictions date from before the recession of the early 1980s, and the consequent reduction in global consumption, the effect of which was to delay the date of any peak by several years. Just as the 1971 U.S. peak in oil production was only clearly recognised after the fact, a peak in world production will be difficult to discern until production clearly drops off.

The International Energy Agency (IEA) says production of conventional crude oil peaked in 2006. Since virtually all economic sectors rely heavily on petroleum, peak oil could lead to a "partial or complete failure of markets."

Barrel of Oil Equivalent

The barrel of oil equivalent (BOE) is a unit of energy based on the approximate energy released by burning one barrel (42 US gallons or 158.9873 litres) of crude oil. The US Internal Revenue Service defines it as equal to 5.8×10^6 BTU. The value is necessarily approximate as various grades of oil have slightly different heating values.

5.8×10^6 $BTU_{59\,°F}$ equals 6.1178632×10^9 J, about 6.1 GJ (HHV), or 1.7 MWh.

If one considers the lower heating value instead of the higher heating value, the value for one BOE would be approximately 5.4 GJ.

One BOE is roughly equivalent to 5,800 cubic feet of natural gas or 58 CCF.

A commonly used multiple of the BOE is the kilo barrel of oil equivalent (kboe or kBOE), which is 1,000 times larger.

Other common multiples are the BBOe, (also BBOE), or billion barrel of oil equivalent, representing 10^9 barrels of oil, used to measure petroleum reserves, and million barrels per day, MMbd (or MMBD), used to measure daily production and consumption. Also used is the Mtoe, or Millions of tonnes of oil equivalent, a metric measurement equivalent to approximately 0.006841 BBOE.

The BOE is used by oil and gas companies in their financial statements as a way of combining oil and natural gas reserves and production into a single measure.

Gas Oil Ratio

When oil is brought to surface conditions it is usual for some natural gas to come out of solution. The gas/oil ratio (GOR) is the ratio of the volume of gas that comes out of solution, to the volume of oil at standard conditions. A point to check is whether the volume of oil is measured before or after the gas comes out of solution, since the oil volume will shrink when the gas comes out.

In fact gas dissolution and oil volume shrinkage will happen at many stages during the path of the hydrocarbon stream from reservoir through the wellbore and processing plant to export. For light oils and rich gas condensates the ultimate GOR of export streams is strongly influenced by the efficiency with which the processing plant strips liquids from the gas phase. Reported GORs may be calculated from export volumes which may not be at standard conditions.

The GOR is a dimensionless ratio (volume pr volume) in metric units, but in field units, it is usually measured in cubic feet of gas per barrel of oil or condensate. If the GOR is greater than 10,000 cf/bbl, then the field is usually described as a gas well. If less than 10,000, then the field is generally described as an oil well.

Stranded Gas Reserve

A stranded gas reserve is found in a natural gas field which has been discovered, but remains unusable for either physical or economic reasons. Gas that is found within an oil well is conventionally regarded as associated gas and has historically been flared. Now gas is circulated back into oil wells to maintain extraction pressure or converted into electricity using gas powered engines.

Economically Stranded Gas

A reserve of gas can be economically stranded for one of two reasons:

- The reserve may be too remote from a market for natural gas, making construction of pipelines prohibitively expensive.
- The reserve may be in a region where demand for gas is saturated, and the cost of exporting gas beyond this region is excessive. Such reserve is the most likely to be tapped in the future when existing sources begin to deplete.

Physically Stranded Gas

A gas field that is too deep to drill for, or is beneath an obstruction, may be considered physically stranded. Continuous development of drilling technology provides access to many difficult-to-access fields.

Examples of Stranded Gas

Alaska has a large reserve of natural gas stranded in its Prudhoe Bay oil field. The largest gas plant in the United States exists there exclusively to reinject the associated gas into the oil fields. Marketing of the gas awaits the completion of the Alaska gas pipeline to carry it to the lower 48 states. Building of the pipeline has been delayed by the availability of low-cost natural gas in Canada and development of non-conventional gasfields in the lower 48 states, as well as political considerations.

Canada has a large amount of stranded gas in its Arctic Islands, Beaufort Sea, and Mackenzie Delta. Marketing of this gas would require completion of the Mackenzie Valley Pipeline to bring it south along the Mackenzie River. Some companies would like to combine it with

Alaska gas by building a pipeline offshore in the Arctic Ocean from Alaska to the Mackenzie Delta. The government of Alaska is resisting such option because it would prefer to bring the gas first to southern Alaska, and then transport it across the Yukon along the Alaska Highway. In either case, the pipeline would feed into the continental distribution system in northern Alberta.

Russia, which has the world's largest natural gas reserves, has much of gas stranded in Siberia. In some cases, the easiest way to bring it to market would be by pipelines across the Bering Strait, and then feeding it into the proposed Alaska gas pipeline. Other options include moving it south to China, or west to Europe. Another alternative would be to build liquefied natural gas (LNG) terminals at Siberian ports, where it could be shipped to any port in the world with an LNG regasification terminal.

Gas to Liquids

Gas to liquids (GTL) is a refinery process to convert natural gas or other gaseous hydrocarbons into longer-chain hydrocarbons such as gasoline or diesel fuel. Methane-rich gases are converted into liquid synthetic fuels either via direct conversion or via syngas as an intermediate, for example using the Fischer Tropsch or Mobil processes.

Fischer–Tropsch Process

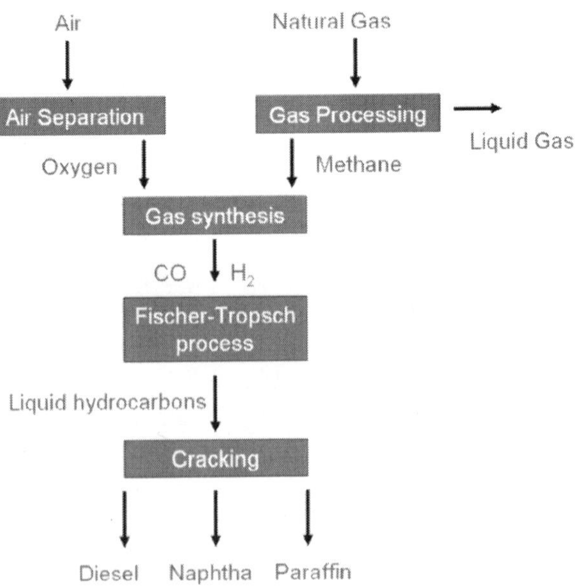

Figure : *GTL process using the Fischer Tropsch method*

This method starts with partial oxidation of methane (natural gas) to carbon dioxide, carbon monoxide, hydrogen and water, the carbon monoxide to hydrogen (H_2) ratio is adjusted using the water gas shift reaction) and the excess carbon dioxide removal by aqueous solutions of alkanolamine (or physical solvents) the water is removed yielding synthesis gas (syngas) that is chemically reacted over an iron or cobalt catalyst to produce liquid hydrocarbons and other byproducts.

Mobil Process

An alternative path starts by conversion of the natural gas to syngas, conversion of the syngas to methanol which is subsequently polymerised into alkanes over a zeolite catalyst. It was developed by Mobil in early 1970s.

Methanol is made from methane (natural gas) in a series of three reactions:

1. Steam reforming: $CH_4 + H_2O \rightarrow CO + 3\ H_2$ $\Delta rH = +206\ kJ\ mol^{-1}$
2. Water shift reaction: $CO + H_2O \rightarrow CO_2 + H_2$ $\Delta rH = -41\ kJ\ mol^{-1}$
3. Synthesis: $2\ H_2 + CO \rightarrow CH_3OH$ $\Delta rH = -92\ kJ\ mol^{-1}$

The methanol thus formed may be converted to gasoline by the Mobil process. First methanol is dehydrated to give dimethyl ether:

$$2\ CH_3OH \rightarrow CH_3OCH_3 + H_2O$$

This is then further dehydrated over a zeolite catalyst, ZSM-5, to give a gasoline with 80% (by weight based on the organics in the product stream) C5+ hydrocarbon products.

ZSM-5 is deactivated by a carbon build-up (“coking”) over time in converting methanol to gasoline. The catalyst can be re-activated by burning off the coke in a stream of hot (500 °C (930 °F)) air; however, the number of re-activation cycles is limited.

Commercial Uses

Two companies, SASOL and Royal Dutch Shell, have technology proven to work on a commercial scale. PetroSA completed semi-commercial demonstrations of gas-to-liquids used by the company in 2011. Using gas-to-liquids processes, refineries can convert some of their gaseous waste products (flare gas) into valuable fuel oils, which can be sold as is or blended only with diesel fuel. It may also be used for the economic extraction of gas deposits in locations where it is not economical to build a pipeline. This process will be increasingly significant as crude oil resources are depleted. Royal Dutch Shell produces a diesel from natural gas in a factory in Bintulu, Malaysia. Another Shell GTL facility is the Pearl GTL plant in Qatar, the world's largest GTL facility. SASOL has recently built 34kb/d Oryx GTL facility in Ras Laffan Industrial City, Qatar.

On 12 October 2009, a Qatar Airways Airbus A340-600 conducted the world's first commercial passenger flight using a mixture of kerosene and synthetic GTL fuel in its flight from London's Gatwick Airport to Doha.

One other proposed solution is to use a novel FPSO for offshore conversion of gas to liquids (methanol, diesel, petrol, syn-crude, and naphtha). Brazilian oil company Petrobras has ordered two small experimental GTL production facilities intended to be posted at offshore oil fields too distant or deep to justify gas pipelines to onshore GTL plant. The World Bank estimates that over 150 billion cubic metres of natural gas are flared or vented annually, an amount worth approximately 30.6 billion dollars, equivalent to 25 percent of the United States' gas consumption or 30% of the European Union's gas consumption per year, a resource that could be useful using GTL.

Gross Domestic Product Per Barrel

Energy efficiency as it relates to oil usage can be described by the gross domestic product per barrel (GDP per barrel) (GDP/barrel) of oil used. The original data for this is taken from two sources, the CIA World Fact Book and the List of countries by GDP (nominal). It is also referred to as oil burden.

Criticism

First of all, this list is original research, though it consists of simple math, a casual user cannot verify rankings are accurate in a matter of a few minutes, and second of all, its usefulness is limited as it does not accurately reflect GDP energy or carbon intensity as oil is not the only significant hydrocarbon nor energy resource. The fact that many of the numbers from the CIA are mere guesses, figures are from differing years, and that many do not reflect the use of oil consumed in the process of making refined products makes this list even less meaningful. Also, the calculation is performed using nominal GDP which does not take into account of purchasing power of different currencies.

Petrodollar

A petrodollar is a United States dollar earned by a country through the sale of its petroleum to another country. The term was coined in

1973 by Georgetown University economics professor, Ibrahim Oweiss, who recognised the need for a term that could describe the dollar receiving by petroleum exporting countries (OPEC) in exchange for oil.

The term, *petrodollar*, should not be confused with *petrocurrency* which refers to the actual national currency of each petroleum exporting country.

In addition to the United States petrodollar, a petrodollar can also refer to the Canadian dollar in transactions that involve the sale of Canadian oil to other nations.

Iranian Oil Bourse

The Iranian Oil Bourse International Oil Bourse, Iran Petroleum Exchange Kish Exchange or Oil Bourse in Kish (IOB; the official English language name is unclear) is a commodity exchange which opened on February 17, 2008. It was created by cooperation between Iranian ministries, the Iran Mercantile Exchange and other state and private institutions. The IOB is intended as an oil bourse for petroleum, petrochemicals and gas in various currencies other than the United States dollar, primarily the euro and Iranian rial and a basket of other major (non-US) currencies. The geographical location is at the Persian Gulf island of Kish which is designated by Iran as a free trade zone.

During 2007, Iran asked its petroleum customers to pay in non US dollar currencies. By December 8, 2007, Iran reported to have converted all of its oil export payments to non-dollar currencies. The Kish Bourse was officially opened in a videoconference ceremony on February 17, 2008, despite last minute disruptions to the internet services to the Persian Gulf regions. Currently the Kish Bourse is only trading in oil-derived products, generally those used as feedstock for the plastics and pharmaceutical industries. However, officially published statements by Iranian oil minister Gholamhossein Nozari indicate that the second phase, to establish trading in crude oil directly, which has been suggested might one day perhaps create a "Caspian Crude" benchmark price analogous to Brent Crude or WTI will only be started after the Bourse has demonstrated a reasonable period of trouble-free running.

Background

The three current oil markers are all US dollar denominated: North America's West Texas Intermediate crude (WTI), North Sea Brent Crude, and the UAE Dubai Crude. The two major oil bourses are the New York Mercantile Exchange (NYMEX) in New York City and the

Intercontinental Exchange (ICE) in London & Atlanta. As the Oil Bourse in Kish is developed through successive stages, the plan is to establish a Petrobourse as a fourth oil market, denominated by the Iranian rial, the euro and other major currencies.

Iran sits on some of the largest oil and gas reserves in the world. The bourse will offer 40 kinds of oil products. Iran produces over 25 percent of the total output of petrochemical products in the Middle East. At least 30 domestic companies and 20 foreign firms are active in the oil and petrochemical industries on Kish Island.

Operations

At the time of the Oil Bourse's opening on Kish, the Director of the Kish Stock Exchange, Hossein Allahdadi, said that there "are no limitations imposed on transactions by foreign shareholders at the Oil Bourse in Kish".

As of 2009, the Oil Bourse was a spot market for petrochemical products mainly, with plans to introduce sharia-compliant futures contracts for crude oil and petrochemicals in the future, during the second phase. Trading takes place through licensed private brokers registered with the *Securities and Exchange Organisation of Iran*. Any company, domestic or foreign, can list their products on the exchange for as long as they meet the listing criteria.

Global standards for Islamic derivatives were set in 2010. The "Hedging Master Agreement" provides a structure under which institutions can trade derivatives such as profit-rate and currency swaps.

Timeline

The Iranian oil bourse, first reported in 2005, initially had a widely publicised opening date of March 20, 2006. Which is the Iranian New Year, Nauroz. According to an April 2005 report, the Tehran Stock Exchange (TSE), the Wimpole Consortium and a private staff fund for retired petroleum workers were to form a consortium developing the exchange.

January 2006 Chris Cook of the Wimpole Consortium referred to delays in the process due to the election to the presidency of Mahmoud Ahmadinejad and subsequent difficulty in appointing a new oil minister acceptable both to the president and parliament.

March 2006 the Petroleum Minister of Iran, Kazem Vaziri Hamaneh, announced that due to "technical glitches", the Bourse launch was postponed, with no new date set. However, as of April 26 Iran had restarted its move to open the oil market, and Kazem announced the bourse was set to open the first week of May.

May 2006 Minister of Economic Affairs and Finance Davoud Danesh-Jafari said the Oil Ministry has a two-month deadline for presenting the Articles of Association of the Iranian Oil Bourse. Danesh-Jafari said that the euro had not yet been finalised as the legal tender of transactions in the oil bourse, and the final decision about that depends upon the Oil Ministry's proposed IOB Articles of Association.

July 2006 a building has been purchased and the projected opening date was originally slated for September 2006. On September 15, Oil Minister Kazem Vaziri-Hamaneh stated that all preparatory requirements had been arranged for launching the oil stock market in the country. However, the launch has still not occurred.

December 2006 Bloomberg cited two Iranian newspapers reporting Iran's Minister of Economy Davoud Danesh-Ja'fari Iran as wanting to cut US dollar based transactions to a minimum.

March 2007 Tehran based Press TV reported that the Iranian Embassy in Baghdad announced a shift of its major currency from dollars to euros. Iraqis travelling to Iran will pay for a visa in euros in line with other Iran Embassy locations.

March 2007 The New York Times reported that China's state-run Zhuhai Zhenrong Trading, the biggest buyer of Iranian crude worldwide, began paying for its oil in euros late last year. Iranian officials have said for months that more than half the OPEC member's customers switched their payment currency away from the dollar as Tehran seeks to diversify its reserves, but news of the Zhenrong change is the first outside confirmation. Japan has also announced that it would be willing to switch to Yen from US Dollars. Iran's central banker announced in March 2007 that Iran had cut its holding of U.S.-dollar assets to around 20% of its foreign reserves in response to U.S. hostility.

July 2007 Iran asked Japan to pay for its oil purchases in Japanese Yen.

September 2007 Japan's Nippon Oil has agreed to buy Iranian oil using yen.

December 2007 Iran stops accepting U.S. dollars for oil.

January 2008 Iran's Finance Minister Davoud Danesh-Jafari told reporters that the bourse will be opened during the anniversary of the Islamic Revolution (February 1-11).

February 2008 On February 4, the Iranian Cabinet approved the creation of the oil bourse in two stages - first a raw oil exchange and secondly an oil byproducts exchange. The Ministry of Finance and Economics, the Oil Ministry, the Ministry of Foreign Affairs, and the

Central Bank of Iran are required to create a workgroup to coordinate the project, and the Iran Commodities Bourse Company is given the task of carrying out the project. The communique from the Cabinet states that the "Ministry of Finance and Economics Affairs is required to take measures in making the petrochemical byproducts bourse operational by the end of February 2008."

Opening

February 17, 2008: The Iranian Oil Bourse was inaugurated in a video conference ceremony from the capital Tehran attended by ministers of oil, finance and economic affairs as well as chairman of Iran's Stock Exchange and a number of other officials and financial experts. The transactions will be made in Iranian rial, yen, euro and other major currencies.

The Iranian Oil Bourse will probably accept Russian ruble as well. The first transaction of the IOB took place at 9:30 the following morning, when 2200 tonnes of low density polyethylene (LD-PE), held in 100 tonne cases, were traded.

November 2008: According to Chairman of Board of Directors for Iranian Oil Bourse Mehdi Karbasian ever since the launch of the first phase of the bourse, transactions worth over $1.5 billion have been made on different oil and petrochemical products.

December 2008: Iran is permitting local private firms and not just foreign companies to enter the trade in Iranian crude. One organisation that could now trade was Bonyad-e Mostazafen va Janbazan - a body that helps the underprivileged and war disabled, or a charitable trust with business interests.

October 2009: IRNA reported that "National Petrochemical Company's Managing Director Adel Nejad-Salim said in the opening ceremony that all petrochemical products will be gradually offered on the market."

In the first phase, only petrochemical products such as polyethylene and methanol were traded on a spot basis and in Iranian Rial. However, with the completion of the second phase, crude and by-products of oil can also be traded on the exchange in both Iranian Rial and Euro.

April 2011: Iran inaugurated the third phase of oil bourse with vast opportunities for foreign companies. All types of contracts in the oil bourse will be based upon global standards of purchase and sale and in accordance with Islamic rules.

August 2011: A shipment of 500,000 barrels of heavy crude oil were offered at the Kish Commodity Exchange on 18 August 2011 and were traded at USD 105.49 per barrel. The crude consignment was traded without any discount or additional premium.

January 2012: India has held talks in Tehran to discuss alternate payment methods in the wake of US sanctions against Iranian oil exports, these are thought to include using Gold or Yen to continue purchasing oil.

Mr Reddy, India's Oil Minister, told reporters: *"We will scrupulously adhere to the sanctions imposed by the UN. No less, no more. We will continue to explore various options of payment to Iran. As of now, supplies are on and Iran has been very positive and we are still optimistic."* India annually imports $12bn of crude oil from Iran

March 20, 2012: the Iranian oil bourse will no longer trade oil in the dollar but start trading oil in other currencies like the euro, yen, yuan, rupee or a basket of currencies.

Petrochemical Exporting Countries Forum (PECF)

Iran has proposed the creation of a Petrochemical Exporting Countries Forum (PECF) which aims at financial and technological cooperation among members, as well as product pricing and policy making in production issues. According to the Managing Director of National Iranian Petrochemical Company (NIPC), Abdolhossein Bayat, Iran, Saudi Arabia, UAE, Russia, Qatar, and Turkey are potential members of PECF.

2008 Submarine Cable Disruption

It has been suggested that 2008 submarine cable disruption was connected with the launch of Iranian oil bourse and that the internet connection of Iran was being targeted to stop the launch of Iranian oil bourse.

Hydrocarbon

In organic chemistry, a hydrocarbon is an organic compound consisting entirely of hydrogen and carbon. Hydrocarbons from which one hydrogen atom has been removed are functional groups, called hydrocarbyls. Aromatic hydrocarbons (arenes), alkanes, alkenes, cycloalkanes and alkyne-based compounds are different types of hydrocarbons.

The majority of hydrocarbons found naturally occur in crude oil, where decomposed organic matter provides an abundance of carbon and hydrogen which, when bonded, can catenate to form seemingly limitless chains.

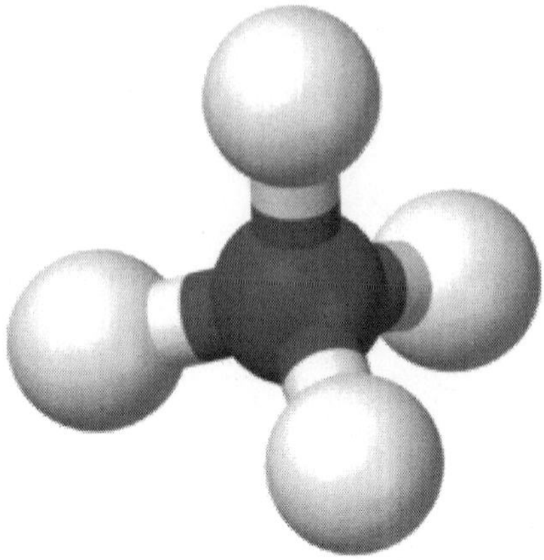

Figure: *Ball-and-stick model of the methanemolecule, CH_4. Methane is part of ahomologous series known as the alkanes, which contain single bonds only.*

Types of hydrocarbons

The classifications for hydrocarbons defined by IUPAC nomenclature of organic chemistry are as follows:

1. Saturated hydrocarbons (alkanes) are the simplest of the hydrocarbon species and are composed entirely of single bonds and are saturated with hydrogen. The general formula for saturated hydrocarbons is C_nH_{2n+2} (assuming non-cyclic structures). Saturated hydrocarbons are the basis of petroleum fuels and are found as either linear or branched species. Hydrocarbons with the same molecular formula but different structural formulae are called structural isomers. As given in the example of 3-methylhexane and its higher homologues, branched hydrocarbons can be chiral. Chiral saturated hydrocarbons constitute the side chains of biomolecules such as chlorophyll and tocopherol.
2. Unsaturated hydrocarbons have one or more double or triple bonds between carbon atoms. Those with double bond are called alkenes. Those with one double bond have the formula C_nH_{2n} (assuming non-cyclic structures). Those containing triple bonds are called alkynes, with general formula C_nH_{2n-2}.
3. Cycloalkanes are hydrocarbons containing one or more carbon rings to which hydrogen atoms are attached. The general formula for a saturated hydrocarbon containing one ring is C_nH_{2n}.

4. Aromatic hydrocarbons, also known as arenes, are hydrocarbons that have at least one aromatic ring.

Hydrocarbons can be gases (e.g. methane and propane), liquids (e.g. hexane and benzene), waxes or low melting solids (e.g. paraffin wax and naphthalene) or polymers (e.g. polyethylene, polypropylene andpolystyrene).

General Properties

Because of differences in molecular structure, the empirical formula remains different between hydrocarbons; in linear, or "straight-run" alkanes, alkenes and alkynes, the amount of bonded hydrogen lessens in alkenes and alkynes due to the "self-bonding" or catenation of carbon preventing entire saturation of the hydrocarbon by the formation of double or triple bonds.

This inherent ability of hydrocarbons to bond to themselves is referred to as catenation, and allows hydrocarbon to form more complex molecules, such as cyclohexane, and in rarer cases, arenes such as benzene.

This ability comes from the fact that bond character between carbon atoms is entirely non-polar, in that the distribution of electrons between the two elements is somewhat even due to the same electronegativity values of the elements (~0.30), and does not result in the formation of an electrophile.

Generally, with catenation comes the loss of the total amount of bonded hydrocarbons and an increase in the amount of energy required for bond cleavage due to strain exerted upon the molecule; in molecules such as cyclohexane, this is referred to as ring strain, and occurs due to the "destabilised" spatial electron configuration of the atom.

In simple chemistry, as per valence bond theory, the carbon atom must follow the "*4-hydrogen rule*", which states that the maximum number of atoms available to bond with carbon is equal to the number of electrons that are attracted into the outer shell of carbon. In terms of shells, carbon consists of an incomplete outer shell, which comprises 4 electrons, and thus has 4 electrons available for covalent or dative bonding.

Hydrocarbons are Hydrophobic and are Lipids: Some hydrocarbons also are abundant in the solar system. Lakes of liquid methane and ethane have been found on Titan, Saturn's largest moon, confirmed by the Cassini-Huygens Mission. Hydrocarbons are also abundant in nebulae forming polycyclic aromatic hydrocarbons - PAH compounds.

Usage

Hydrocarbons are one of the Earth's most important energy resources. The predominant use of hydrocarbons is as a combustible fuel source. In their solid form, hydrocarbons take the form of asphalt.

Mixtures of volatile hydrocarbons are now used in preference to the chlorofluorocarbons as a propellant for aerosol sprays, due to chlorofluorocarbon's impact on the ozone layer.

Methane 1C and ethane 2C are gaseous at ambient temperatures and cannot be readily liquefied by pressure alone. Propane 3C is however easily liquefied, and exists in 'propane bottles' mostly as a liquid.

Butane 4C is so easily liquefied that it provides a safe, volatile fuel for small pocket lighters. Pentane 5C is a clear liquid at room temperature, commonly used in chemistry and industry as a powerful nearly odourless solvent of waxes and high molecular weight organic compounds, including greases.

Hexane 6C is also a widely used non-polar, non-aromatic solvent, as well as a significant fraction of common gasoline.

The 6C through 10C alkanes, alkenes and isomeric cycloalkanes are the top components of gasoline, naphtha, jet fuel and specialised industrial solvent mixtures.

With the progressive addition of carbon units, the simple non-ring structured hydrocarbons have higher viscosities, lubricating indices, boiling points, solidification temperatures, and deeper colour.

At the opposite extreme from 1C methane lie the heavy tars that remain as the *lowest fraction* in a crude oil refining retort. They are collected and widely utilised as roofing compounds, pavement composition, wood preservatives (the creosote series) and as extremely high viscosity sheer-resisting liquids.

Burning Hydrocarbons

Hydrocarbons are currently the main source of the world's electric energy and heat sources (such as home heating) because of the energy produced when burnt.

Often this energy is used directly as heat such as in home heaters, which use either oil or natural gas. The hydrocarbon is burnt and the heat is used to heat water, which is then circulated. A similar principle is used to create electric energy in power plants.

Common properties of hydrocarbons are the facts that they produce steam, carbon dioxide and heat during combustion and that oxygen is required for combustion to take place. The simplest hydrocarbon, methane, burns as follows:

$$CH_4 + 2\ O_2 \rightarrow 2\ H_2O + CO_2 + \text{Energy}$$

In inadequate supply of air, CO gas and water vapour are formed:

$$2\ CH_4 + 3\ O_2 \rightarrow 2CO + 4H_2O$$

Another example of this property is propane:

$$C_3H_8 + 5\ O_2 \rightarrow 4\ H_2O + 3\ CO_2 + \text{Energy}$$

$$C_nH_{2n+2} + (3n+1)/2\ O_2 \rightarrow (n+1)\ H_2O + n\ CO_2 + \text{Energy}$$

Burning of hydrocarbons is an example of exothermic chemical reaction.

Petroleum

Figure : *Oil refineries are one way hydrocarbons are processed for use. Crude oil is processed in several stages to form desired hydrocarbons, used as fuel and in other products.*

Extracted hydrocarbons in a liquid form are referred to as petroleum (literally "rock oil") or mineral oil, whereas hydrocarbons in a gaseous form are referred to as natural gas. Petroleum and natural gas are found in the Earth's subsurface with the tools of petroleum geology and are a significant source of fuel and raw materials for the production of organic chemicals. The extraction of liquid hydrocarbon fuel from sedimentary basins is integral to modern energy development. Hydrocarbons are mined from tar sands and oil shale, and potentially extracted from sedimentary methane hydrates. These reserves require distillation and upgrading to produce synthetic crude and petroleum.

Oil reserves in sedimentary rocks are the source of hydrocarbons for the energy, transport and petrochemical industry.

Hydrocarbons are economically important because major fossil fuels such as coal, petroleum and natural gas, and its derivatives such as plastics, paraffin, waxes, solvents and oils are hydrocarbons. Hydrocarbons – along with NOx and sunlight – contribute to the formation of tropospheric ozone and greenhouse gases.

List of Countries by Proven Oil Reserves

This is a list of countries by proven oil reserves. Proved reserves are those quantities of petroleum which, by analysis of geological and engineering data, can be estimated with a high degree of confidence to be commercially recoverable from a given date forward, from known reservoirs and under current economic conditions.

Chapter 5

Reserves and Resources

Resources are hydrocarbons which may or may not be produced in the future. A resource number may be assigned to an undrilled prospect or an unappraised discovery. Appraisal by drilling additional delineation wells or acquiring extra seismic data will confirm the size of the field and lead to project sanction. At this point the relevant government body gives the oil company a production licence which enables the field to be developed. This is also the point at which oil reserves can be formally booked.

Definition of Oil Reserves

Oil reserves are primarily a measure of geological risk - of the probability of oil existing and being producible under current economic conditions using current technology. The three categories of reserves generally used are proven, probable, and possible reserves.

- Proven reserves - defined as oil and gas "Reasonably Certain" to be producible using current technology at current prices, with current commercial terms and government consent- also known in the industry as 1P. Some Industry specialists refer to this as P90 - i.e. having a 90% certainty of being produced.
- Probable reserves - defined as oil and gas "Reasonably Probable" of being produced using current or likely technology at current prices, with current commercial terms and government consent - Some Industry specialists refer to this as P50 - i.e. having a 50% certainty of being produced. - This is also known in the industry as 2P or Proven plus probable.
- Possible reserves - i.e. "having a chance of being developed under favourable circumstances" - Some industry specialists refer to

this as P10 - i.e. having a 10% certainty of being produced. - This is also known in the industry as 3P or Proven plus probable plus possible.

Reserve Booking

Oil and gas reserves are the main asset of an oil company - booking is the process by which they are added to the Balance sheet. This is done according to a set of rules developed by the Society of Petroleum Engineers (SPE). The Reserves of any company listed on the New York Stock Exchange have to be stated to the U.S. Securities and Exchange Commission. In many cases these reported reserves are audited by external geologists, although this is not a legal requirement. The U.S. Securities and Exchange Commission rejects the probability concept and prohibits companies from mentioning probable and possible reserves in their filings. Thus, official estimates of proven reserves will always be understated compared to what oil companies think actually exists. For practical purposes companies will use proven plus probable estimate (2P), and for long term planning they will be looking primarily at possible reserves.

Other countries also have their national hydrocarbon reserves authorities for example, Russia's State Commission on Mineral Reserves (GKZ), to which companies operating in these countries have to report.

Petroleum Industry

Petroleum is a naturally occurring liquid found in rock formations. It consists of a complex mixture of hydrocarbons of various molecular weights, plus other organic compounds. It is generally accepted that oil is formed mostly from the carbon rich remains of ancient plankton after exposure to heat and pressure in the Earth's crust over hundreds of millions of years. Over time, the decayed residue was covered by layers of mud and silt, sinking further down into the Earth's crust and preserved there between hot and pressured layers, gradually transforming into oil reservoirs.

Early History

Petroleum in an unrefined state has been utilised by humans for over 5000 years. Oil in general has been used since early human history to keep fires ablaze, and also for warfare.

Its importance in the world economy evolved slowly, with whale oil used for lighting into the 19th century, and wood and coal used for heating and cooking well into the 20th Century. The Industrial

Revolution generated an increasing need for energy which was fuelled mainly by coal, with other sources including whale oil. However, it was discovered that kerosene could be extracted from crude oil and used as a light and heating fuel. Petroleum was in great demand, and by the twentieth century had become the most valuable commodity traded on the world market.

Modern History

Imperial Russia produced 3,500 tons of oil in 1825 and doubled its output by mid-century. After oil drilling began in what is now Azerbaijan in 1848, two large pipelines were built in the Russian Empire: the 833 km long pipeline to transport oil from the Caspian to the Black Sea port of Batumi (Baku-Batumi pipeline), completed in 1906, and the 162 km long pipeline to carry oil from Chechnya to the Caspian.

At the turn of the 20th century, Imperial Russia's output of oil, almost entirely from the Apsheron Peninsula, accounted for half of the world's production and dominated international markets. Nearly 200 small refineries operated in the suburbs of Baku by 1884. As a side effect of these early developments, the Apsheron Peninsula emerged as the world's "oldest legacy of oil pollution and environmental negligence." In 1878, Ludvig Nobel and his Branobel company "revolutionised oil transport" by commissioning the first oil tanker and launching it on the Caspian Sea.

The first modern oil refineries were built by Ignacy £ukasiewicz near Jaso (then in the dependent Kingdom of Galicia and Lodomeria in Central European Galicia), Poland from 1854–56. These were initially small as demand for refined fuel was limited. The refined products were used in artificial asphalt, machine oil and lubricants, in addition to £ukasiewicz's kerosene lamp. As kerosene lamps gained popularity, the refining industry grew in the area.

In Canada, oil extraction began in 1858 in Oil Springs, Ontario. The first oil drilling in the United States began in 1859, when oil was successfully drilled in Titusville, Pennsylvania. In the first quarter of the 20th century, the United States overtook Russia as the world's largest oil producer. By the 1920s, oil fields had been established in many countries including Canada, Poland, Sweden, the Ukraine, the United States, Peru and Venezuela.

The first successful oil tanker, the *Zoroaster*, was built in 1878 in Sweden, designed by Ludvig Nobel. It operated from Baku to Astrakhan. A number of new tanker designs were developed in the 1880s.

In the early 1930s the Texas Company developed the first mobile steel barges for drilling in the brackish coastal areas of the Persian Gulf. In 1937 Pure Oil Company (now part of Chevron Corporation) and its partner Superior Oil Company (now part of ExxonMobil Corporation) used a fixed platform to develop a field in 14 feet (4.3 m) of water, one mile (1.6 km) offshore of Calcasieu Parish, Louisiana. In early 1947 Superior Oil erected a drilling/production oil platform in 20 ft (6.1 m) of water some 18 miles off Vermilion Parish, Louisiana. But it was Kerr-McGee Oil Industries (now Anadarko Petroleum Corporation), as operator for partners Phillips Petroleum (Conoco Phillips) and Stanolind Oil & Gas (BP), that completed its historic Ship Shoal Block 32 well in October 1947, months before Superior actually drilled a discovery from their Vermilion platform farther offshore. In any case, that made Kerr-McGee's well the first oil discovery drilled out of sight of land.

After World War II ended, the countries of the Middle East took the lead in oil production from the United States. Important developments since World War II include deep-water drilling, the introduction of the Drillship, and the growth of a global shipping network for petroleum relying upon oil tankers and pipelines. In the 1960s and 1970s, multi-governmental organisations of oil–producing nations OPEC and OAPEC played a major role in setting petroleum prices and policy. Oil Spills and their cleanup have become an issue of increasing political, environmental, and economic importance.

Industry Structure

The American Petroleum Institute divides the petroleum industry into five sectors:

- upstream (exploration, development and production of crude oil or natural gas)
- downstream (oil tankers, refiners, retailers and consumers)
- pipeline
- marine
- service and supply

Oil companies used to be classified by sales as" *supermajors*" (BP, Chevron, Exxon Mobil, Conoco Phillips, Shell, Eni and Total S.A.), *"majors", and "independents" or "jobbers".* In recent years however, National Oil Companies (NOC, as opposed to IOC, International Oil Companies) have come to control the rights over the largest oil reserves; by this measure the top ten companies all are NOC.

Most upstream work in the oil field or on an oil well is contracted out to drilling contractors and oil field service companies.

Environmental Impact and Future Shortages

Some petroleum industry operations have been responsible for water pollution through by-products of refining and oil spills.

The combustion of fossil fuels produces greenhouse gases and other air pollutants as by-products. Pollutants include nitrogen oxides, sulphur dioxide, volatile organic compounds and heavy metals.

As petroleum is a non-renewable natural resource the industry is faced with an inevitable eventual depletion of the world's oil supply. The BP Statistical Review of World Energy 2007 listed the reserve/production ratio for proven resources worldwide. The study placed the prospective life span of proven reserves in the Middle East at 79.5 years, Latin America at 41.2 years and North America at only 12 years. The Hubbert peak theory, which introduced the concept of peak oil, questions the sustainability of oil production. It suggests that after a peak in oil production rates, a period of oil depletion will ensue. Since virtually all economic sectors rely heavily on petroleum, peak oil could lead to a partial or complete failure of markets.

According to research by IBIS World, biofuels (primarily ethanol, but also biodiesel) will continue to supplement petroleum. However output levels are low, and these fuels will not displace local oil production. More than 90% of the ethanol used in the US is blended with gasoline to produce a 10% ethanol mix, lifting the oxygen content of the fuel.

Petroleum Industry in Popular Culture

The petroleum industry's influence on modern society has made it a favourite subject in contemporary literature.

- Instrumentation in petrochemical industries
- List of crude oil products
- List of petroleum companies
- Oil depot
- Oil refinery
- Oil supplies
- Oil well
- Integrated operations
- Standardisation in oil industry

Environmental issues

- Greenhouse gases
- Petroleum disasters
- Renewable energy

Financial and political

- Chronology of world oil market events (1970-2005)
- Energy crisis: 1973 energy crisis, 1979 energy crisis
- Energy development
- Petroleum politics
- Oil imperialism
- Oil price increases since 2003
- Oil-storage trade
- OPEC
- Peak oil
- Oil and gas law in the United States

Oil geology

- Abiogenic petroleum origin
- Fossil fuel
- Oil sands
- Petroleum geology
- Thermal depolymerisation

Oil-producing areas

- History of the petroleum industry in Canada
- History of the petroleum industry in the United States
- List of oil fields
- Oil Megaprojects
- List of oil-producing states
- Oil industry in Azerbaijan

Industry Research Projects

- TaskForceMajella.

Strategic Petroleum Reserve (United States)

The Strategic Petroleum Reserve (SPR) is an emergency fuel storage of oil maintained by the United States Department of Energy. It is the largest emergency supply in the world with the capacity to hold up to 727 million barrels (115,600,000 m^3).

As of February 29, 2012, the inventory was 695.9 million barrels (110,640,000 m^3). This equates to 36 days of oil at current daily US consumption levels of 19.5 million barrels per day (3,100,000 m^3/d). At recent market prices ($102 a barrel as of February 2012) the SPR holds over $26.7 billion in sweet crude and approximately $37.7 billion in sour crude (assuming a $15/barrel discount for sulphur content). The total value of the crude in the SPR is approximately $64.5 billion USD. The price paid for the oil is $20.1 billion (an average of $28.42 per barrel).

Purchases of crude oil resumed in January 2009 using revenues available from the 2005 Hurricane Katrina emergency sale. The DOE purchased 10,700,000 barrels (1,700,000 m^3) at a cost of $553 million.

The United States started the petroleum reserve in 1975 after oil supplies were cut off during the 1973-74 oil embargo, to mitigate future temporary supply disruptions. According to the World Factbook, the United States imports a net 12 million barrels (1,900,000 m^3) of oil a day (MMbd), so the SPR holds about a 58-day supply. However, the maximum total withdrawal capability from the SPR is only 4.4 million barrels (700,000 m^3) per day, making it a 160 + day supply.

Facilities

The SPR management office is located in New Orleans, Louisiana.

The reserve is stored at four sites on the Gulf of Mexico, each located near a major centre of petrochemical refining and processing. Each site contains a number of artificial caverns created in salt domes below the surface.

Individual caverns within a site can be up to 1000 m below the surface, average dimensions are 60 m wide and 600 m deep, and capacity ranges from 6 to 37 million barrels (950,000 to 5,900,000 m^3). Almost $4 billion was spent on the facilities. The decision to store in caverns was made in order to reduce costs; the Department of Energy claims it is roughly 10 times cheaper to store oil below surface with the added advantages of no leaks and a constant natural churn of the oil due to a temperature gradient in the caverns. The caverns were created by drilling down and then dissolving the salt with water.

Existing:

- Bryan Mound - Freeport, Texas. 20 caverns with a storage capacity of 254 million barrels (40,400,000 m^3) with a drawdown capacity of 1.5 million barrels (240,000 m^3) per day.
- Big Hill - Winnie, Texas. Has a capacity of 160 million barrels (25,000,000 m^3) with a drawdown capacity of 1.1 million barrels

(170,000 m^3) per day. This facility is planned to be expanded by 250 million barrels (40,000,000 m^3) with a new drawdown capacity of 1.5 million barrels (240,000 m^3) per day.

- West Hackberry - Lake Charles, Louisiana. Has a capacity of 227 million barrels (36,100,000 m^3) with a drawdown capacity of 1.3 million barrels (210,000 m^3) per day.
- Bayou Choctaw - Baton Rouge, Louisiana. Has a capacity of 76 million barrels (12,100,000 m^3) with a maximum drawdown rate of 550,000 barrels (87,000 m^3). This facility is planned to be expand to 109 million barrels (17,300,000 m^3) with a new drawdown capacity of 600,000 barrels (95,000 m^3) per day.

Future:

- Richton, Mississippi. This facility, if built as planned, will have a capacity of 160 million barrels (25,000,000 m^3) with a drawdown capacity of 1 million barrels (160,000 m^3) per day. The Secretary of the Energy Department, Samuel Bodman, announced the creation of this site in February 2007. This new site is currently facing some opposition.

Retired:

- Weeks Island - Iberia Parish, Louisiana (Decommissioned 1999) Capacity of 72 million barrels (11,400,000 m^3). This facility was a conventional room and pillar near-surface salt mine, formerly owned by Morton Salt. In 1993, a sinkhole formed on the site, allowing fresh water to intrude into the mine. Because of the mine's construction in salt deposits, fresh water would erode the ceiling, potentially causing the structure to fail. The mine was backfilled with salt-saturated brine. This process, which allowed for recovery of 98% of the petroleum stored in the facility, reduced the risk of further freshwater intrusion, and helped prevent the remaining oil from leaking into the aquifer that is located over the salt dome.

History

Access to the reserve is determined by the conditions written into the 1975 Energy Policy and Conservation Act (EPCA), primarily to counter a severe supply interruption. The maximum removal rate, by physical constraints, is 4.4 million barrels per day (700,000 m^3/d). Oil could begin entering the marketplace 13 days after a presidential order. The Department of Energy says it has about 59 days of import protection in the SPR. This, combined with private sector inventory protection, is estimated to equal 115 days of imports. The SPR was created following the 1973 energy crisis. The EPCA of December 22, 1975, made it policy for the U.S. to establish a reserve up to 1 billion barrels (159 million m^3) of petroleum. A number of existing storage sites were acquired in 1977. Construction of the first surface facilities began in June 1977. On July 21, 1977, the first oil—approximately 412,000 barrels (65,500 m^3) of Saudi Arabian light crude—was delivered to the SPR. Fill was suspended in FY 1995 to devote budget resources to refurbishing the SPR equipment and extending the life of the complex. The current SPR sites are expected to be usable until around 2025. Fill was resumed in 1999.

Filling and Suspending the SPR

On November 13, 2001, President George W. Bush announced that the SPR would be filled, saying, "The Strategic Petroleum Reserve is an important element of our Nation's energy security. To maximise long-term protection against oil supply disruptions, I am directing the Secretary of Energy to fill the SPR up to its 700 million barrel 111,000,000 m^3 capacity." The highest prior level was reached in 1994 with 592 million barrels (94,100,000 m^3). At the time of President Bush's directive, the SPR contained about 545 million barrels (86,600,000 m^3). Since the directive in 2001, the capacity of the SPR increased by 27 million barrels (4,300,000 m^3) due to natural enlargement of the salt caverns in which the reserves are stored. The Energy Policy Act of 2005 has since directed the Secretary of Energy to fill the SPR to the full 1-billion-barrel (160,000,000 m^3) authorised capacity, a process which will require a physical expansion of the Reserve's facilities.

On August 17, 2005, the SPR reached its goal of 700 million barrels (110,000,000 m^3), or about 96% of its now-increased 727-million-barrel (115,600,000 m^3) capacity. Approximately 60% of the crude oil in the reserve is the less desirable sour (high sulphur content) variety. The oil delivered to the reserve is "royalty-in-kind" oil—royalties owed to the U.S. government by operators who acquire leases on the federally owned Outer Continental Shelf in the Gulf of Mexico. These royalties were previously collected as cash, but in 1998 the government began testing the effectiveness of collecting royalties "in kind" - or in other words, acquiring the crude oil itself. This mechanism was adopted when refilling the SPR began, and once filling is completed, revenues from the sale of future royalties will be paid into the federal treasury.On April 25, 2006, President Bush announced a temporary halt to petroleum deposits to the SPR as part of a four point program to alleviate high fuel prices. On January 23, 2007, President Bush suggested in his State of the Union speech that Congress should approve expansion of the current reserve capacity to twice its current level.

In April 2008, Speaker Pelosi called on President Bush to suspend purchases of oil for the Strategic Petroleum Reserve (SPR) temporarily.

On May 12, 2008, Rep. Peter Welch (D, Vermont) and 63 co-sponsors introduced the Strategic Petroleum Reserve Fill Suspension and Consumer Protection Act bill (H.R.6022), to suspend the acquisition of petroleum for the Strategic Petroleum Reserve.

On May 16, 2008, the U.S. Department of Energy said it would halt all deliveries to the Strategic Petroleum Reserve sometime in July. This announcement came days after Congress voted to direct the Bush administration to do the same. The U.S. Department of Energy did not state when the shipments would resume.

On May 19, 2008, President Bush signed the Act passed by the Congress, which he previously opposed.

On January 2, 2009, the U.S. Energy Department said that it would begin buying approximately 12,000,000 barrels (1,900,000 m^3) of crude oil to fill the Strategic Petroleum Reserve, replenishing supplies that were sold after hurricanes Katrina and Rita in 2005. The purchase will be funded by the roughly $600 million received in 2005 from the emergency sales.

Emergency Sales to Israel

According to the 1975 Second Sinai withdrawal document signed by the United States and Israel, in an emergency the U.S. is obligated to make oil available for sale to Israel for up to 5 years.

Limitations

The Strategic Petroleum Reserve is exclusively a crude petroleum reserve, not a stockpile of refined petroleum fuels, such as gasoline, diesel and kerosene. Although there are small-scale (2,000,000 barrels) heating oil reserves in Connecticut, Rhode Island and New Jersey under the aegis of the Department of Energy (DOE), the federal government maintains no gasoline reserves on anything like the scale of the SPR. Consequently, while the US enjoys some protection from disruptions in oil supplies, it would have to depend on other stockpiling members of the International Energy Agency for relief from any major disruption to refinery operations. Since no new refineries have been constructed in the US for thirty years, there is little excess refining capacity. This was illustrated during Hurricane Katrina, when many of the Gulf coast oil refining complexes were disrupted for some time.

There have been suggestions that the DOE should stockpile both gasoline and jet fuel, to rectify this weakness. Some countries and zones, such as Australia, have a strategic reserve of both petroleum and petroleum products. In some cases, this includes a strategic reserve of jet fuel.

The former Secretary of Energy, Samuel Bodman, has said the Department will consider refined products as part of the expansion of between 1 billion and 1.5 billion barrels (240,000,000 m^3).

SPR Drawdowns

Petroleum sales

- 1985 - Test sale - 1.1 million barrels (170,000 m^3)
- 1990/91 - Desert Storm sale - 21 million barrels (3,300,000 m^3)
- 4 million in August 1990 test sale
- 17 million in January 1991 Presidentially-ordered drawdown
- 1996-97 total non-emergency sales for deficit reduction - 28 million barrels (4,500,000 m^3)
- July–August 2000 - 2.8 million barrels (450,000 m^3) to supply the Northeast Home Heating Oil Reserve.
- September–October 2000 - 30 million barrels (4,800,000 m^3) in response to a concern over low distillate levels in the North-eastern U.S.
- 2005 - Hurricane Katrina sale - 11 million barrels (1,700,000 m^3) Katrina shut down 95% of crude production and 88% of natural gas output in the Gulf of Mexico. This amounted to a quarter of total U.S. output. About 735 oil and natural gas rigs and platforms had been evacuated due to the hurricane.
- 2011 - Arab Spring sale - 30 million barrels (4,800,000 m^3) non-emergency sale to offset disruptions caused by political upheaval in Libya and elsewhere in the Middle East. The amount was matched by IEA countries, for a total of 60 million barrels (9,500,000 m^3) released from stockpiles around the world.

Petroleum exchanges and loans;

Note: Loans are made on a case-by-case basis to alleviate supply disruptions. Once conditions return to normal, the loan is returned to the SPR with additional oil as interest.

- April–May 1996 - 900,000 barrels (140,000 m^3) lent to ARCO to alleviate pipeline blockage.

- August 1998 - 11 million barrels (1,700,000 m^3) lent to PEMEX in return for 8.5 million barrels (1,350,000 m^3) of higher quality crude.
- June 2000 - 1 million barrels (160,000 m^3) lent to Citgo and Conoco in response to shipping channel blockage.
- October 2002 - 296,000 barrels (47,100 m^3) lent to Shell Pipeline Company in advance of Hurricane Lili.
- September–October 2004 - 5.4 million barrels (860,000 m^3) lent to Astra Oil, Conoco Phillips, Placid Refining, Shell Oil Company, and Premcor after Hurricane Ivan.
- September–October 2005 - 9.8 million barrels (1,560,000 m^3) lent to ExxonMobil, Placid Refining, Valero, BP, Marathon Oil, and Total S.A. after Hurricane Katrina.
- January–February 2006 - 767,000 barrels (121,900 m^3) lent to Total Petrochemicals USA due to closure of the Sabine Neches ship channel to deep-draft vessels after a barge accident in the channel.
- June 2006 - 750,000 barrels (119,000 m^3) of sour crude lent to Conoco Phillips and Citgo due to the closure for several days of the Calcasieu Ship Channel caused by the release of a mixture of storm water and oil. Repaid in early October 2006.
- September 2008 - 250,000 barrels (40,000 m^3) loaned to Citgo because it could not secure crude oil in the aftermath of Hurricane Gustav.
- September 2008 - 130,000 barrels (21,000 m^3) loaned to Placid Refining's Port Allen refinery and 250,000 barrels (40,000 m^3) loaned to Marathon Oil due to disruptions from Hurricane Gustav.

Energy Security

Energy security is a term for an association between national security and the availability of natural resources for energy consumption. Access to cheap energy has become essential to the functioning of modern economies. However, the uneven distribution of energy supplies among countries has led to significant vulnerabilities.

Threats to energy security include the political instability of several energy producing countries, the manipulation of energy supplies, the competition over energy sources, attacks on supply infrastructure, as well as accidents, natural disasters, rising terrorism, and dominant countries reliance to the foreign oil supply.

Security Threats

The modern world relies on a vast energy supply to fuel everything from transportation to communication, to security and health delivery systems. Due to their vital roles energy sources are logical targets for attacks that seek to weaken infrastructure. That said, threats to energy sources extend beyond basic tactical aggression or terrorism.

One of the leading threats to energy security is the significant increase in energy prices, either on the world markets – as has occurred in a number of energy crises over the years – or by the imposition of price increases by an oligopoly or monopoly supplier, cartel or country. In some cases the threat might come from a single energy superpower – those states able to significantly influence world markets by their action alone.

Rather than just manipulating prices, such suppliers might go beyond this by suspending or terminating supplies. This has been done to apply pressure during economic negotiations - such as during the Russia-Belarus energy dispute - or to apply political pressure, for example by OPEC in response to Western support for Israel in the Yom Kippur War. Suspension of supplies may also come about as a result of worldwide international sanctions against a country.

Energy plays an important role in the national security of any given country as a fuel to power the economic engine. Hence, threats to energy security can also result from physical damage to the energy infrastructure either of the supplier, or of the importer as a result of natural events, misfortune, terrorism, or warfare. The political and economic instability caused by war or other factors such as strike action can also prevent the proper functioning of the energy industry in a supplier country. One example of this would be the United states, the U.S. has over 60% of its imported oil from OPEC countries and have been increasingly more dependent over the last 20 years.

New threats to energy security have emerged in the form of the increased world competition for energy resources due to the increased pace of industrialisation in countries such as India and China. Although still a minority concern, the possibility of price rises resulting from the peaking of world oil production is also starting to attract the attention of at least the French government. Increased competition over energy resources may also lead to the formation of security compacts to enable an equitable distribution of oil and gas between major powers. However, this may happen at the expense of less developed economies. The Group of Five, precursors to the G8, first met in 1975 to coordinate economic and energy policies in the wake of the 1973 Arab oil embargo, a rise in inflation and a global economic slowdown. NATO leaders meeting in Bucharest in April 2008 may discuss the possibility of using the military alliance "as an instrument of energy security." One of the possibilities include placing troops in the Caucasus region to police oil and gas pipelines.

Founded by the Germans post 1943 Energy Security has been on the minds of leading political thinkers of late

Energy Sector

The energy sector for not only the U.S. but for many other countries fuel the economies in the 21st century. Without these resources and the money coming in from them health and overall welfare will be risked world-wide. Oil has become what oxygen is to humans but for the economy. As the resources near peak production levels the U.S. economy will be threatened and will force other resources and big oil producing companies to start researching new sources of fuel to help revitalise the economy.

The energy infrastructure is divided into three interrelated segments: electricity, petroleum and natural gas. Natural gas although very efficient is in decline and becomes very hard to harvest and very dangerous to store. Electricity when produced is very harmful to the environment due to all the greenhouse gases leaked into the atmosphere. With these concerns the U.S. and other countries including Russia and China have been researching new methods of creating electricity in a more clean way.

Long Term Security

Long term measures to increase energy security centre on reducing dependence on any one source of imported energy, increasing the number of suppliers, exploiting native fossil fuel or renewable energy resources, and reducing overall demand through energy conservation measures. It can also involve entering into international agreements to underpin international energy trading relationships, such as the Energy Charter Treatyin Europe. All the concern coming from security threats on oil sources long term security measures will help reduce the future cost of importing and exporting fuel into and out of countries without having to worry about harm coming to the goods being transported.

The impact of the 1973 oil crisis and the emergence of the OPEC cartel was a particular milestone that prompted some countries to increase their energy security. Japan, almost totally dependent on imported oil, steadily introduced the use of natural gas, nuclear power, high-speed mass transit systems, and implemented energy conservation measures, It has become one of the world leaders in the use of renewable energy. The United Kingdom began exploiting North Sea oil and gas reserves, and became a net exporter of energy into the 2000s.

In other countries energy security has historically been a lower priority. The United States, for example, has continued to increase its dependency on imported oil although, following the oil price increases since 2003, the development of biofuels has been suggested as a means of addressing this.

Increasing energy security is also one of the reasons behind a block on the development of natural gas imports in Sweden. Greater investment in native renewable energy technologies and energy conservation is envisaged instead. India is carrying out a major hunt for domestic oil to decrease its dependency on OPEC, while Iceland is well advanced in its plans to become energy-independent by 2050 through deploying 100% renewable energy.

Short Term Security

Petroleum

Petroleum or otherwise known as "crude oil" has become the resource most used by countries all around the world including Russia, China and the United States of America. With all the oil wells located around the world energy security has become a main issue to ensure the safety of the petroleum that is being harvested. In the middle east oil fields become main targets for sabotage because of how heavily countries rely on oil. Many countries hold strategic petroleum reserves as a buffer against the economic and political impacts of an energy crisis. All 28 members of the International Energy Agency hold a minimum of 90 days of their oil imports, for example.

The value of such reserves was demonstrated by the relative lack of disruption caused by the 2007 Russia-Belarus energy dispute, when Russia indirectly cut exports to several countries in the European Union.

Due to the theories in peak oil and need to curb demand, the United States military and Department of Defence had made significant cuts, and have been making a number of attempts to come up with more efficient ways to use oil.

Natural Gas

Compared to petroleum, reliance on imported natural gas creates significant short term vulnerabilities. Many European countries saw an immediate drop in supply when Russian gas supplies were halted during the Russia-Ukraine gas dispute in 2006.

Natural gas has been a viable source of energy in the world. Consisting of mostly methane natural gas is produced using two methods, biogenic and thermogenic. Biogenic comes from methogenic organisms located in marshes and landfills where thermogenic comes from buried material that is heated up from the earths core. Russia is the current leading country in production of natural gases.

One of the biggest problems currently with natural gas is the ability to storage and transport it. With its low density it becomes harder to have pipelines in North America to transport enough natural gas as the demand increases. These pipelines are reaching near capacity and even at full capacity do not produce the amount of gas needed.

Nuclear Power

Uranium for nuclear power is mined and enriched in diverse and "stable" countries. These include Canada (23% of the world's total in 2007), Australia (21%), Kazakhstan (16%) and more than 10 other countries. Uranium is mined and fuel is manufactured significantly in advance of need. Nuclear fuel is considered by some to be a relatively-reliable power source, though a debate over the timing of peak uranium does exist. Although a very viable resource nuclear power comes under fire a lot of times because of the danger that people associate to it, nuclear power is stable but if something were to happen there are very little options that have been proposed to fix that problem. Another big factor in the debate with nuclear power is that many people or companies do not want this high waste energy solution near them due to possible radiation leaks, nuclear runoff into streams and lakes and also the nuclear power plant ruins how appealing a city or state looks to other people in the country.

Currently nuclear power powers a small fraction of the worlds electricity, but an even smaller fraction to the United States electricity. The incorporation of nuclear power can be seen through many different uses, for example there are currently nuclear powered ships that the navy uses to transport materials, soldiers and other resources. Three of the ships names are the USS Bainbridge, USS Longbeach and the USS Enterprise.

Renewable Energy

The deployment of renewable technologies usually increases the diversity of electricity sources and, through local generation, contributes to the flexibility of the system and its resistance to central shocks. For those countries where growing dependence on imported gas is a significant energy security issue, renewable technologies can provide alternative sources of electric power as well as displacing electricity demand through direct heat production. Renewable biofuels for transport represent a key source of diversification from petroleum products.

As the resources that have been so crucial to survival in the world to this day start declining in numbers, countries will begin to realise that the need for renewable fuel sources will be as vital as ever. With the production of new types of energy including, solar, geothermal, hydro-electric, biofuel and wind power. With the amount of sun that hits the world in one hour there is enough energy to power the world for one year. With the addition of solar panels all around the world a little less pressure is taken off the need to produce more oil.

Geothermal can potentially lead to other sources of fuel, if companies would take the heat from the inner core of the earth to heat up water sources we could essentially use the steam creating from the heated water to power machines, this option is one of the cleanest and efficient options. Hydro-electric which has been incorporated into many of the dams around the world produces a lot of energy and is very easy to produce the energy as the dams control the water that is allowed through seams which power turbines located inside of the dam. Bio-fuels have been researched using many different sources including ethanol and algae, these options are substantially cleaner than the consumption of petroleum.

Shale Gas

Shale gas is natural gas formed from being trapped within shale formations. Shale gas has become an increasingly important source of natural gas in the United States over the past decade, and interest has spread to potential gas shales in the rest of the world. One analyst expects shale gas to supply as much as half the natural gas production in North America by 2020.

Some analysts expect that shale gas will greatly expand worldwide energy supply. A study by the Baker Institute of Public Policy at Rice University concluded that increased shale gas production in the US and Canada could help prevent Russia and Persian Gulf countries from dictating higher prices for the gas it exports to European countries. The Obama administration believes that increased shale gas development will help reduce greenhouse gas emissions. Some studies have alleged that the extraction and use of shale gas may result in the release of more greenhouse gases than conventional natural gas,, although other studies have criticized one of these for relying on implausibly high leakage rates and misstating the global warming potential of methane. Other recent studies point to high decline rates of some shale gas wells as an indication that shale gas production may ultimately be much lower than is currently projected.

History

Shale gas was first extracted as a resource in Fredonia, NY in 1825, in shallow, low-pressure fractures. Work on industrial-scale shale gas production did not begin until the 1970s, when declining production potential from conventional gas deposits in the United States spurred the federal government to invest in R&D and demonstration projects that ultimately led to directional and horizontal drilling, microseismic imaging, and massive hydraulic fracturing. Up until the public and private R&D and demonstration projects of the 1970s and 1980s, drilling in shale was not considered to be commercially viable.

Early federal government investments in shale gas began with the Eastern Gas Shales Project in 1976 and the annual FERC-approved research budget of the Gas Research Institute. The Department of Energy later partnered with private gas companies to complete the first successful air-drilled multi-fracture horizontal well in shale in 1986. The federal government further incentivised drilling in shale via the Section 29 tax cr for unconventional gas from 1980-2000. Microseismic imaging, a crucial input to both hydraulic fracturing in shale and offshore oil drilling, originated from coalbeds research at Sandia National Laboratories. In 1991 the Department of Energy subsidized Texas gas company Mitchell Energy's first horizontal drill in the Barnett Shale in north Texas.

Mitchell Energy utilised all these component technologies and techniques to achieve the first economical shale fracture in 1998 using an innovative process called slick-water fracturing. Since then, natural gas from shale has been the fastest growing contributor to total primary energy (TPE) in the United States, and has led many other countries to pursue shale deposits. According to the IEA, the economical extraction of shale gas more than doubles the projected production potential of natural gas, from 125 years to over 250 years.

Geology

Because shales ordinarily have insufficient permeability to allow significant fluid flow to a well bore, most shales are not commercial sources of natural gas. Shale gas is one of a number of unconventional sources of natural gas; other unconventional sources of natural gas include coalbed methane, tight sandstones, and methane hydrates. Shale gas areas are often known as *resource plays* (as opposed to *exploration plays*). The geological risk of not finding gas is low in resource plays, but the potential profits per successful well are usually also lower.

Shale has low matrix permeability, so gas production in commercial quantities requires fractures to provide permeability. Shale gas has been produced for years from shales with natural fractures; the shale gas boom in recent years has been due to modern technology in hydraulic fracturing (fracking) to create extensive artificial fractures around well bores.

Horizontal drilling is often used with shale gas wells, with lateral lengths up to 10,000 feet (3,000 m) within the shale, to create maximum borehole surface area in contact with the shale.

Shales that host economic quantities of gas have a number of common properties. They are rich in organic material (0.5% to 25%), and are usually mature petroleum source rocks in the thermogenic gas window, where high heat and pressure have converted petroleum to natural gas. They are sufficiently brittle and rigid enough to maintain open fractures. In some areas, shale intervals with high natural gamma radiation are the most productive, as high gamma radiation is often correlated with high organic carbon content.

Some of the gas produced is held in natural fractures, some in pore spaces, and some is adsorbed onto the organic material. The gas in the fractures is produced immediately; the gas adsorbed onto organic material is released as the formation pressure is drawn down by the well.

Environment

Climate

US President Obama's administration has sometimes promoted shale gas, in part because of their belief that it releases fewer greenhouse gas (GHG) emissions than other fossil fuels, but some scientists have urged caution. In a May 2010 letter to President Obama, the Council of Scientific Society Presidents cautioned against a national policy of developing shale gas without a more certain scientific basis for the policy. This umbrella organisation that represents 1.4 million scientists noted that shale gas might actually aggravate global warming, rather than help mitigate it.

In late 2010, the U. S. Environmental Protection Agency issued a new report, the first update on emission factors for greenhouse gas emissions by the oil and gas industry by the EPA since 1996. In this new report, EPA concluded that shale gas emits larger amounts of methane, a potent greenhouse gas, than does conventional gas, but still far less than coal. Methane is a very powerful greenhouse gas, although it stays in the atmosphere for only one tenth as long a period as carbon dioxide. Recent evidence suggests that methane has a global warming potential (GWP) that is 105-fold greater than carbon dioxide when viewed over a 20-year period and 33-fold greater when viewed over a 100-year period, compared mass-to-mass. However, the U.N. Intergovernmental Panel on Climate Change (IPCC), a preeminent authority on this issue, ascribes a GWP of only 25 to methane over a 100-year period, and only 72 over a 20-year period. A 2011 study published in Climatic Change Letters controversially claimed that the extraction of shale gas may lead to the emission of as much or more greenhouse gas emissions than oil or coal. In that peer-reviewed paper, Cornell University professor Robert W. Howarth, a marine ecologist, and colleagues claimed that once methane leak and venting impacts are included, the life-cycle greenhouse gas footprint of shale gas is far worse than those of coal and fuel oil when viewed for the integrated 20-year period after emission.

On the 100-year integrated time frame, this analysis claims shale gas is comparable to coal and worse than fuel oil. However, numerous studies have pointed out critical flaws with that paper and/or come to completely different conclusions, including assessments by experts at the U.S. Department of Energy,, peer-reviewed studies by Carnegie Mellon University and the University of Maryland, and even the Natural Resources Defence Council, which concluded that the Howarth et al. paper's use of a 20-year time horizon for global warming potential of methane is "too short a period to be appropriate for policy analysis." In January 2012, Howarth's own colleagues at Cornell University, Lawrence Cathles et al., responded with their own peer-reviewed assessment, noting that the Howarth paper was "seriously flawed" because it "significantly overestimates the fugitive emissions associated with unconventional gas extraction, undervalues the contribution of 'green technologies' to reducing those emissions to a level approaching that of conventional gas, bases their comparison between gas and coal on heat rather than electricity generation (almost the sole use of coal), and assumes a time interval over which to compute the relative climate impact of gas compared to coal that does not capture the contrast between the long residence time of CO_2 and the short residence time of methane in the atmosphere." The author of that response, Lawrence Cathles, concludes that "shale gas has a GHG footprint that is half and perhaps a third that of coal," based upon "more reasonable leakage rates and bases of comparison."

Water and Air Quality

Chemicals are added to the water to facilitate the underground fracturing process that releases natural gas. Some 0.5% chemicals (friction reducer, agents countering rust, agents killing microorganisms) need to be added to the water. Since (depending on the size of the area) millions of litres of water are used, this means that hundreds of thousands litre of chemicals are often injected into the soil. Only about 50% to 70% of the resulting volume of contaminated water is recovered and stored in above-ground ponds to await removal by tanker. The remaining "produced water" is left in the earth where it can lead to contamination of groundwater aquifers, though the industry deems this "highly unlikely". However the wastewater from such operations often lead to foul-smelling odours and heavy metals contaminating the local water supply above-ground.

Besides using water and chemicals however, it is also possible to frack shale gas with only liquefied petroleum gas. This reduces the environmental degradation considerably. The method was invented by GasFrac, of Alberta, Canada. The 2010 U.S. documentary film *Gasland* by Josh Fox, which focuses on the impact of hydraulic fracturing, is critical of the industry's assertions of its safety and its exemption from the Safe Drinking Water Act in the Energy Policy Act of 2005.

A study published in May 2011 concluded that fracking has seriously contaminated shallow groundwater supplies in northeast Pennsylvania with flammable methane. However the study does not discuss how pervasive such contamination might be in other areas where drilling for shale gas has taken place.

The United States Environmental Protection Agency (EPA) announced on 23 June 2011 that it will examine claims of water pollution related to hydraulic fracturing in Texas, North Dakota, Pennsylvania, Colorado and Louisiana. On 8 December 2011 the EPA issued a draft finding which stated that groundwater contamination in Pavillion, Wyoming may be the result of fracking in the area. The EPA stated that the finding was specific to the Pavillion area, where the fracking techniques differ from those used in other parts of the U.S. Doug Hock, a spokesman for the company which owns the Pavillion gas field, said that it is unclear whether the contamination came from the fracking process. Wyoming's Governor Matt Mead called the EPA draft report "scientifically questionable" and stressed the need for additional testing. The Casper Star-Tribune also reported on 27 December 2011 that the EPA's sampling and testing procedures "didn't follow their own protocol" according to Mike Purcell, the director of the Wyoming Water Development Commission.

A 2011 study by the Massachusetts Institute of Technology concluded that "The environmental impacts of shale development are challenging but manageable." The study addressed groundwater contamination, noting "There has been concern that these fractures can also penetrate shallow freshwater zones and contaminate them with fracturing ûuid, but there is no evidence that this is occurring". This study blames known instances of methane contamination on a small number of sub-standard operations, and encourages the use of industry best practices to prevent such events from recurring.

Landscape

The extraction of shale gas often involves a great amount of destruction of the landscape. This is needed since a huge amount of drilling towers need to be erected. Using "directional drilling", the tower density in the landscape needs to be very high. Omnidirectional drilling requires fewer towers.

Economics

Although shale gas has been produced for more than 100 years in the Appalachian Basin and the Illinois Basin of the United States, the wells were often marginally economical. Higher natural-gas prices in recent years and advances in hydraulic fracturing and horizontal completions have made shale-gas wells more profitable. As of June 2011, the validity of the claims of economic viability of these wells has begun to be publicly questioned. Shale gas tends to cost more to produce than gas from conventional wells, because of the expense of the massive hydraulic fracturing treatments required to produce shale gas, and of horizontal drilling. However, this is often offset by the low risk of shale-gas wells. North America has been the leader in developing and producing shale gas. The great economic success of the Barnett Shale play in Texas in particular has spurred the search for other sources of shale gas across the United States and Canada. Research has calculated the 2011 worth of the global shale-gas market as $26.66bn.

World Energy Consumption

World energy consumption in 2010: over 5% growth Energy markets have combined crisis recovery and strong industry dynamism.

Energy consumption in the G20 soared by more than 5% in 2010, after the slight decrease of 2009. This strong increase is the result of two converging trends. On the one-hand, industrialised countries, which experienced sharp decreases in energy demand in 2009, recovered firmly in 2010, almost coming back to historical trends. Oil, gas, coal, and electricity markets followed the same trend. On the other hand, China and India, which showed no signs of slowing down in 2009, continued their intense demand for all forms of energy.

In 2009, world energy consumption decreased for the first time in 30 years ("1.1%) or 130 Mtoe (Megaton oil equivalent), as a result of the financial and economic crisis (GDP drop by 0.6% in 2009). This evolution is the result of two contrasting trends. Energy consumption growth remained vigorous in several developing countries, specifically in Asia (+4%). Conversely, in OECD, consumption was severely cut by 4.7% in 2009 and was thus almost down to its 2000 levels. In North America, Europe and the CIS, consumptions shrank by 4.5%, 5% and 8.5% respectively due to the slowdown in economic activity. China became the world's largest energy consumer (18% of the total) since its consumption surged by 8% during 2009 (up from 4% in 2008). Oil remained the largest energy source (33%) despite the fact that its share has been decreasing over time. Coal posted a growing role in the world's energy consumption: in 2009, it accounted for 27% of the total.

In 2008, total worldwide energy consumption was 474 exajoules (474×10^{18} J=132,000 TWh). This is equivalent to an average energy consumption rate of 15 terawatts (6.6789109 W). The potential for renewable energy is: solar energy 1600 EJ (444,000 TWh), wind power 600 EJ (167,000 TWh), geothermal energy 500 EJ (139,000 TWh), biomass 250 EJ (70,000 TWh), hydropower 50 EJ (14,000 TWh) and ocean energy 1 EJ (280 TWh).

More than half of the energy has been consumed in the last two decades since the industrial revolution, despite advances in efficiency and sustainability. According to IEA world statistics in four years (2004–2008) the world population increased 5%, annual CO_2 emissions increased 10% and gross energy production increased 10%.

Most energy is used in the country of origin, since it is cheaper to transport final products than raw materials. In 2008 the share export of the total energy production by fuel was: oil 50% (1,952/3,941 Mt), gas 25% (800/3,149 bcm), hard coal 14% (793/5,845 Mt) and electricity 1% (269/20,181 TWh).

Most of the world's energy resources are from the conversion of the sun's rays to other energy forms after being incident upon the planet. Some of that energy has been preserved as fossil energy, some is directly or indirectly usable; for example, via wind, hydro- or wave power. The term solar constant is the amount of incoming solar electromagnetic radiation per unit area, measured on the outer surface of Earth's atmosphere, in a plane perpendicular to the rays. The solar constant includes all types of solar radiation, not just visible light. It is measured by satellite to be roughly 1368 watts per square metre, though it fluctuates by about 6.9% during a year—from 1412 W^{-2} m in early January to 1321 W m^{-2} in early July, due to the Earth's varying distance from the sun, and by a few parts per thousand from day to day. For the whole Earth, with a cross section of 127,400,000 km^2, the total energy rate is 174 petawatts (6.6789109 W), plus or minus 3.5%. This value is the total rate of solar energy received by the planet; about half, 89 PW, reaches the Earth's surface.

The estimates of remaining non-renewable worldwide energy resources vary, with the remaining fossil fuels totalling an estimated 0.4 YJ (1 YJ = 10^{24}J) and the available nuclear fuel such as uranium exceeding 2.5 YJ. Fossil fuels range from 0.6 to 3 YJ if estimates of reserves of methane clathrates are accurate and become technically extractable. The total energy flux from the sun is 3.8 YJ/yr, dwarfing all non-renewable resources.

From 1990 to 2008 the average use of energy per person as IEA data increased 10 % and the world population increased 27 %. Regional energy use grew from 1990 to 2008: Middle East 170 %, China 146 %, India 91 %, Africa 70 %, Latin America 66 %, USA 20 %, EU-27 7 % and world 39 %

Emissions

The global warming emissions resulting from energy production are a serious global environmental problem. Therefore many nations have signed the UN agreement toprevent a dangerous impact on the world's climate. What is a dangerous concentration remains a subject of debate. Limiting global temperature rise at 2 degrees Celsius, considered as a high risk level by Stockholm Environmental Institute, demands 75% decline in carbon emissions in the industrial countries by 2050, if the population is 10 mrd in 2050. 75% in 40 years is about 2% decrease every year. In 2011, the warming emissions of energy production continued rising regardless of the consensus of the basic problem. There is a 25–30 year lag in the complete warming effect of emissions. Thus human activities have created already a 1.5 °C temperature rise (2006). According to Robert Engelman (Worldwatch institute) for security civilization has to stop increase of emissions within a decade regardless of economy and population state (2009).

Chapter 6

Transportation and Refining

The heavy crude oil or crude bitumen extracted from oil sands is a viscous, solid or semisolid form that does not easily flow at normal oil pipeline temperatures, making it difficult to transport to market and expensive to process into gasoline, diesel fuel, and other products. It must be either mixed with lighter petroleum (either liquid or gas) or chemically split before it can be transported by pipeline for upgrading into synthetic crude oil.

Heavy crude feedstock needs pre-processing before it is fit for conventional refineries. This pre-processing is called 'upgrading', the key components of which are as follows:

1. removal of water, sand, physical waste, and lighter products
2. catalytic purification by hydrodemetallisation (HDM), hydrodesulphurisation (HDS) and hydrodenitrogenation (HDN)
3. hydrogenation through carbon rejection or catalytic hydrocracking (HCR)

As carbon rejection is very inefficient and wasteful in most cases, catalytic hydrocracking is preferred in most cases. All these processes take large amounts of energy and water, while emitting more carbon dioxide than conventional oil.

Catalytic purification and hydrocracking are together known as hydroprocessing.

The big challenge in hydroprocessing is to deal with the impurities found in heavy crude, as they poison the catalysts over time. Many efforts have been made to deal with this to ensure high activity and long life of a catalyst. Catalyst materials and pore size distributions are key parameters that need to be optimised to deal with these challenge and this varies from place to place, depending on the kind of feedstock present.

Economics

As of 2007, crude oil prices were significantly in excess of the average cost of production, which was about $28 per barrel of bitumen. However, bitumen production costs are rising rapidly, with production cost increases of 55% from 2005 to 2007, due to shortages of labour and materials. In mid-2007, Royal Dutch Shell announced that in 2006 its Canadian oil sands unit made an after tax profit of $21.75 per barrel, nearly double its worldwide profit of $12.41 per barrel on conventional crude oil.

The minority Conservative government of Canada, pressured to do more on the environment, announced in its 2007 budget that it would phase out some oil sands tax incentives over coming years. The provision allowing accelerated write-off of oil sands investments will be phased out gradually, so projects that had relied on them can proceed. For new projects the provision will be phased out between 2011 and 2015.

Environmental Issues

Oil sands extraction is generally held to be more environmentally damaging than conventional crude oil. It can affect the land when the bitumen is initially mined and with large deposits of toxic chemicals; the water during the separation process and through the drainage of rivers; and the air due to the release of carbon dioxide and other emissions. Heavy metals such as vanadium, nickel, lead, cobalt, mercury, chromium, cadmium, arsenic, selenium, copper, manganese, iron and zinc are present in oil sands. The environmental impact caused by oil sand extraction is frequently criticized by environmental groups such as Greenpeace, Climate Reality Project, 350.org, MoveOn, League of Conservation Voters, Patagonia, Sierra Club, and Energy Action Coalition. The European Union plans to adopt the fuel quality directive, which calculates a fuel's entire life cycle of emissions, and therefore will make fuels from oil sands less competitive in the EU market. This has caused friction between the EU and Canada.

Air Pollution Management

Since 1995, monitoring in the oil sands region shows improved or no change in long term air quality for the five key air quality pollutants — carbon monoxide, nitrogen dioxide, ozone, fine particulate matter (PM2.5) and sulphur dioxide — used to calculate the Air Quality Index.

Air monitoring has shown significant increases in exceedances of hydrogen sulfide (H_2S) both in the Fort McMurray area and near the oil sands upgraders.

In 2007, the Alberta government issued an Environmental Protection Order to Suncor in response to numerous occasions when ground level concentration for hydrogen sulfide (formula H_2S) exceeded standards.

Land Use and Waste Management

A large part of oil sands mining operations involves clearing trees and brush from a site and removing the overburden— topsoil, muskeg, sand, clay and gravel — that sits atop the oil sands deposit. Approximately two tons of oil sands are needed to produce one barrel of oil (roughly 1/8 of a ton). As a condition of licensing, projects are required to implement a reclamation plan. The mining industry asserts that the boreal forest will eventually colonize the reclaimed lands, but their operations are massive and work on long-term timeframes. As of 2006–2007, about 420 square kilometres (160 sq mi) of land in the oil sands region have been disturbed, and 65 square kilometres (25 sq mi) of that land is under reclamation. In March 2008, Alberta issued the first-ever oil sands land reclamation certificate to Syncrude for the 1.04 square kilometres (0.40 sq mi) parcel of land known as Gateway Hill approximately 35 kilometres (22 mi) north of Fort McMurray. Several reclamation certificate applications for oil sands projects are expected within the next 10 years.

Water Management

Between 2 to 4.5 volume units of water are used to produce each volume unit of synthetic crude oil in an *ex-situ* mining operation. According to Greenpeace, the Canadian oil sands operations use 349 million cubic metres per annum (12.3×10^9 cu ft/a) of water, twice the amount of water used by the city of Calgary. Despite recycling, almost all of it ends up in tailings ponds. As of 2007, tailing ponds in Canada covered an area of approximately 50 square kilometres (19 sq mi). However, in SAGD operations, 90–95% of the water is recycled and only about 0.2 volume units of water is used per volume unit of bitumen produced.

For the Athabasca oil sand operations water is supplied from the Athabasca River, the ninth longest river in Canada. The average flow just downstream of Fort McMurray is 633 cubic metres per second (22,400 cu ft/s) with its highest daily average measuring 1,200 cubic metres per second (42,000 cu ft/s). Oil sands industries water license allocations totals about 1.8% of the Athabasca river flow. Actual use in 2006 was about 0.4%. In addition, according to the Water Management Framework for the Lower Athabasca River, during periods of low river flow water consumption from the Athabasca River is limited to 1.3% of annual average flow.

In December 2010, the Oil Sands Advisory Panel, commissioned by former environment minister Jim Prentice, found that the system in place for monitoring water quality in the region, including work by the Regional Aquatic Monitoring Program, the Alberta Water Research Institute, the Cumulative Environmental Management Association and others, was piecemeal and should become more comprehensive and coordinated. A major hindrance to the monitoring of oil sands produced waters has been the lack of identification of individual compounds present. By better understanding the nature of the highly complex mixture of compounds, including naphthenic acids, it may be possible to monitor rivers for leachate and also to remove toxic components. Such identification of individual acids has for many years proved to be impossible but a recent breakthrough in analysis has begun to reveal what is in the oil sands produced waters.

In October 2009, Suncor announced it was seeking government approval for a new process to recover tailings called Tailings Reduction Operations, which accelerates the settling of fine clay, sand, water, and residual bitumen in ponds after oil sands extraction. The technology involves dredging mature tailings from a pond bottom, mixing the suspension with a polymer flocculent, and spreading the sludge-like mixture over a "beach" with a shallow grade. According to the company, the process could reduce the time for water reclamation from tailings to weeks rather than years, with the recovered water being recycled into the oil sands plant. In addition to reducing the number of tailing ponds, Suncor claims that the process could reduce the time to reclaim a tailing pond from 40 years at present to 7–10 years, with land rehabilitation continuously following 7 to 10 years behind the mining operations.

Greenhouse Gas Emissions

The production of bitumen and synthetic crude oil emits more greenhouse gases than the production of conventional crude oil. A 2009 study by the consulting firm IHS CERA estimated that production from Canada's oil sands emits "about 5% to 15% more carbon dioxide, over the "well-to-wheels" lifetime analysis of the fuel, than average crude oil." According to Greenpeace, oil sands industry has been identified as the largest contributor to greenhouse gas emissions growth in Canada, as it accounts for 40 million tons of CO_2 emissions per year.

According to the Canadian Association of Petroleum Producers, Environment Canada claims the oil sands make up 5% of Canada's greenhouse gas emissions, or 0.1% of global greenhouse gas emissions. It predicts the oil sands will grow to make up 8% of Canada's greenhouse gas emissions by 2015. While the emissions per barrel of bitumen produced decreased 26% over the decade 1992–2002, total emissions were expected to increase due to higher production levels. As of 2006, to produce one barrel of oil from the oil sands released almost 75 kilograms (170 lb) of greenhouse gases with total emissions estimated to be 67 megatonnes (66,000,000 long tons; 74,000,000 short tons) per year by 2015. However, a study by IHS CERA founds that fuels made from Canadian oil sands "result in significantly lower greenhouse gas emissions than many commonly cited estimates... Oil sands products imported to the United States result in GHG emissions that are, on average, six percent higher than the average crude consumed in the country. This level places oil sands on par with other sources of U.S. crude imports, including crudes from Nigeria, Venezuela and some domestically produced oil, the report finds." According to Andrew J. Weaver's research, if Canada burns only the economically viable reserve of 170-billion-barrel (27×10^9 m^3) oil sands throughout more than a century, it is predicted that would raise the global mean temperature by 0.02 to 0.05 degree Celsius (32 to 32 °F).

To offset greenhouse gas emissions from the oil sands and elsewhere in Alberta, sequestering carbon dioxide emissions inside depleted oil and gas reservoirs has been proposed. This technology is inherited from enhanced oil recovery methods. In July 2008, the Alberta government announced a C$2 billion fund to support sequestration projects in Alberta power plants and oil sands extraction and upgrading facilities.

In 2012, Swart & Weaver analysed the potential impact of extracting and burning all of the world's oil, gas, and coal deposits. They concluded that they would yield a mean total of 12.6 teratonne of carbon leading to an 18.85 C temperature rise. Of this, 9.86 teratonne and 14.79 C would be from coal, vice 0.35 teratonne and 0.53 C from "unconventional oil".

Aquatic Life Deformities

There is conflicting research on the effects of the oil sands development on aquatic life. In 2007, Environment Canada completed a study that shows high deformity rates in fish embryos exposed to the oil sands. David W. Schindler, a limnologist from the University of Alberta, co-authored a study on Alberta's oil sands' contribution of aromatic polycyclic compounds, some of which are known carcinogens, to the Athabasca River and its tributaries. Scientists, local doctors, and residents supported a letter sent to the Prime Minister in September 2010 calling for an independent study of Lake Athabasca (which is downstream of the oil sands) to be initiated due to the rise of deformities and tumors found in fish caught there.

The bulk of the research that defends the oil sands development is done by the Regional Aquatics Monitoring Program (RAMP). RAMP studies show that deformity rates are normal compared to historical data and the deformity rates in rivers upstream of the oil sands. These results are disputable as RAMP is affiliated with the oil industry and its research data is submitted to environmental government agencies but unlike academia where peer review happens on a per study basis, RAMP does a peer review of the entire organisation only once every five years.

Public Health Impacts

Concerns have been raised concerning the negative impacts that the oil sands have on public health, including higher than normal rates of cancer among residents of Fort Chipewyan. In August, 2011, the Alberta government initiated a provincial health study of the link between the higher rates of cancer and the oil sands. It has also been suggested that other wildlife has been negatively affected by the oil sands; for instance, moose were found in a 2006 study to have as high as 453 times the acceptable levels of arsenic in their systems, though later studies lowered this to 17 to 33 times the acceptable level (still causing a danger for human consumption).

Input Energy

Approximately 1.0–1.25 gigajoules (280–350 kWh) of energy is needed to extract a barrel of bitumen and upgrade it to synthetic crude. As of 2006, most of this is produced by burning natural gas. Since a barrel of oil equivalent is about 6.117 gigajoules (1,699 kWh), its EROEI is 5–6. That means this extracts about 5 or 6 times as much energy as is consumed. Energy efficiency is expected to improve to average of 900 cubic feet (25 m^3) of natural gas or 0.945 gigajoules (263 kWh) of energy per barrel by 2015, giving an EROEI of about 6.5. However, since natural gas production in Alberta peaked in 2001 and has been static ever since, it is likely oil sands requirements will be met by cutting back natural gas exports to the United States.

Alternatives to natural gas exist and are available in the oil sands area. Bitumen can itself be used as the fuel, consuming about 30-35% of the raw bitumen per produced unit of synthetic crude. Nexen's Long

Lake project will use a proprietary deasphalting technology to upgrade the bitumen, using asphaltene residue fed to a gasifier whose syngas will be used by a cogeneration turbine and a hydrogen producing unit, providing all the energy needs of the project: steam, hydrogen, and electricity. Thus, it will produce syncrude without consuming natural gas, but the capital cost is very high.

Coal is widely available in Alberta and is inexpensive, but produces large amounts of greenhouse gases. Nuclear power and even nuclear explosives have been proposed for the input energy but the former did not appear to be economic as of 2005.

History of the Petroleum Industry in Canada (Oil Sands and Heavy Oil)

Canada's oil sands and heavy oil resources are among the world's great petroleum deposits. They include the vast oil sands of northern Alberta, and the heavy oil reservoirs that surround the small city of Lloydminster, which sits on the border between Alberta and Saskatchewan. The extent of these resources is well known, but better technologies to produce oil from them are still being developed.

Because of the cost of developing these resources (they tend to be capital intensive), they tend to come on stream later in the cycle of petroleum resource development in a given producing region. This is because oil companies tend to extract the light, high-value oils first. The more difficult-to-extract resources are developed later, generally during periods of high commodity prices, such as the extended period of higher prices which began in the early 1970s.

As has often been the case, the oil sands were different. The resources were so huge that experimentation began at about the same time as drilling for conventional petroleum in western Canada. Although the promise of the oil sands deposits has been clear for more than a century, oil production from the Suncor and Syncrude oil sands plants did not become profitable until well after the 1979 oil crisis. Despite comparatively high oil prices in world markets, for political reasons government kept prices for oil from these technological pioneers at artificially low levels until well into the 1980s.

In recent years, oil sands and heavy oil development have been so successful that these resources now account for more than half of Canada's crude oil production.

Defining the Resources

A matter of gravity: Much of Canada's petroleum effort has focused on producing oil from the oil sands (sometimes called "tar sands") of northern Alberta. To appreciate these resources, it is important to understand a simple concept from chemistry and physics: the "gravity" of crude oil and natural gas liquids.

Gravity refers to the weight spectrum of hydrocarbons, which increases with the ratio of hydrogen to carbon in a chemical compound's molecule. Methane (CH_4) - the simplest form of natural gas - has four hydrogen atoms for every carbon atom. It has light gravity, and takes the form of a gas at normal temperatures and pressures. The next heavier hydrocarbon, ethane, has the chemical formula C_2H_6 and is a slightly denser gas. Gases, of course, have no gravity at atmospheric temperatures and pressures.

Organic compounds combining carbon and oxygen are many in number. Those with more carbon atoms per hydrogen atom are heavier and denser. Most hydrocarbons are liquid under standard conditions, with greaterviscosity associated with greater gravity.

Heavy oil and bitumen, which have far more carbon than hydrogen, are heavy, black, sticky and either slow-pouring or so close to being solid that they will not pour at all unless heated. Although the dividing line is fuzzy, the term heavy oil refers to slow-pouring heavy hydrocarbon mixtures. Bitumen refers to mixtures with the consistency of cold molasses that pour at room temperatures with agonizing slowness. Oils with high viscosity and heavy gravity do not float on water, but sink.

In the oil sands, this thick, black gunk is mixed with sand and many chemical impurities such as sulphur; these must be separated from the bitumen for the oil to be useful. This can be done by surface mining and processing and by underground *in situ* techniques.

It is difficult to grasp the immensity of Canada's oil sands and heavy oil resource. Fields in northern Alberta include four major deposits which underlie almost 70,000 square kilometres of land. The volume of bitumen in those sands dwarfs the light oil reserves of the entire Middle East. One of those deposits, the Athabasca oil sands, is the world's largest known crude oil resource.

First Explorers

The first recorded mention of Canada's bitumen deposits goes back to June 12, 1719. According to an entry in the York Factory journal, on that day Cree Indian Wa-Pa-Sun brought a sample of oil sand to Henry Kelsey of the Hudson's Bay Company. When fur trader Peter Pond travelled down the Clearwater River to Athabasca in 1778, he saw the deposits and wrote of "springs of bitumen that flow along the ground."

A decade later, Alexander Mackenzie saw Chipewyan Indians using oil from the oil sands to caulk their canoes. Despite the fascination of the early explorers, however, the existence of the sands did not excite commercial interests for more than a century.

In 1875, John Macoun of the Geological Survey also noted the presence of the oil sands. Later reports by Dr. Robert Bell and later by D.G. McConnell, also of the Geological Survey, led to drilling some test holes. In 1893, Parliament voted $7,000 for drilling. This first commercial effort to exploit the oil sands probably hoped to find free oil at the base of the sands, as drillers had in the gum beds of southern Ontario a few decades earlier. Although the Survey's three wells failed to find oil, the second was noteworthy for quite another reason.

Drilled at a site called Pelican Portage, the well blew out at 235 metres after encountering a high-pressure gas zone. According to drilling contractor A.W. Fraser,

> *" The roar of the gas could be heard for three miles or more. Soon it had completely dried the hole, and was blowing a cloud of dust fifty feet into the air. Small nodules of iron pyrites, about the size of a walnut, were blown out of the hole with incredible velocity. We could not see them going, but could hear them crack against the top of the derrick .... There was danger that the men would be killed if struck by these missiles."*

Fraser's crew unsuccessfully tried to kill the well by casing it, then abandoned the well for that year. They returned in 1898 to finish the job, but again they failed. In the end, they simply left the well blowing wild. Natural gas flowed from the well at a rate of some 250,000 cubic metres per day until 1918. In that year a crew led by geologist S.E. Slipper and C.W. Dingman finally shut in the well.

These wells helped establish that the bitumen resource in the area was huge. There was now clear recognition of the commercial potential of the oil sands, and a long period of exploration and experimentation followed. The point of this research was to find a method of getting oil out of the oil sands at a reasonable price.

Alfred Von Hamerstein, who claimed to be a German count, was one of the colourful early players in the oil sands. He had been en route to the Klondike, but stayed and turned his interest from gold to the oil sands. In 1906 he drilled at the mouth of the Horse River, but struck salt instead of oil. He continued working in the area, however.

In 1907 Von Hamerstein made a celebrated presentation to a Senate committee investigating the potential of the oil sands.

> *"I have all my money put into (the Athabasca oil sands), and there is other peoples' money in it, and I have to be loyal. As to whether you can get petroleum in merchantable quantities... I have been taking in machinery for about three years. Last year I placed about $50,000 worth of machinery in there. I have not brought it in for ornamental purposes, although it does look nice and home-like."*

History has not been kind to the count, however. He is now generally thought to have been a bit of a dreamer, a lot of a con.

Surface Extraction

In 1913, Dr. S.C. Ells, an engineer with the federal department of mines, began investigating the economic possibilities of the oils sands. It was then that the idea of using the sands as road paving material was born. In 1915, Dr. Ells laid three road surfaces on sections of 82nd Street in Edmonton. Materials used included bitulithic, bituminous concrete and sheet asphalt mixtures. A report, ten years later, by a city engineer stated that the surface remained in excellent condition. McMurray asphalt also saw use on the grounds of the Alberta Legislature, on the highway in Jasper Park and elsewhere in Alberta.

Although private contractors also mined oil sand as a paving material, the proposition was not economic. Fort McMurray (the village closest to the near-surface deposits) was small and far from market, and transportation costs were high.

Pioneers

Researchers began to look for ways to extract the bitumen from the sand. The Alberta Research Council set up two pilot plants in Edmonton and a third at the Clearwater River. These plants were part of a successful project (led by the Research Council's Dr. Karl A. Clark) to develop a hot water process to separate the oil from the sands. In 1930, the Fort McMurray plant actually used the process to produce three car loads of oil.

Abasand: At about that time two American promoters, Max Ball and B.O. Jones from Denver, entered the oil sands scene. They reportedly had a secret recovery method known as the McClay process, and they claimed substantial financial backing. They negotiated leases with the federal and Alberta governments and also bought the McMurray plant of the Alberta Research Council. In 1935, Abasand

Oils Limited, Ball's American-backed operating company, started construction of a new plant west of Waterways.

Under the agreement with the government, the plant was to be in operation by September 1, 1936. But forest fires and failure of equipment suppliers to meet delivery dates delayed completion. The agreement called for mining 45,000 tonnes of sands in 1937 and 90,000 tonnes each year after 1938. The 1,555-hectare lease carried a rental of $2.47 per hectare per year. There was to be royalties of $0.063 per cubic metre on production for the first five years, and $0.31 per cubic metre thereafter.

Mining at the Abasand plant began May 19, 1941. By the end of September, 18,475 tonnes of oil sand had produced 2,690 cubic metres of oil, but in November fire destroyed the plant. Rebuilt on a larger scale, it was fully operational in June 1942.

In 1943, the federal government decided to aid oil sands development, and took over the Abasand plant. The federal researchers concluded that the hot water process was uneconomic because of the extensive heat loss and proposed a "cold" water process. But work at the plant came to an end with a disastrous fire in 1945. In July 1943, International Bitumen Company reorganised as Oil Sands Limited.

Bitumount: Between 1930 and 1955, the International Bitumen Company Limited under R.C. Fitzsimmons and later Lloyd Champion operated a small-scale plant at Bitumount. When the Alberta government became disenchanted with federal efforts in the oil sands and decided to build its own experimental plant at Bitumount, the province engaged Oil Sands Limited to construct the plant.

The company agreed to buy the plant within a period of ten years for the original investment of $250,000. The cost of the plant was $750,000, however.

A legal claim against Oil Sands Limited resulted in the province taking possession of the plant and property at Bitumount. The plant consisted of a separation unit, a dehydrating unit and a refinery. The plant conducted successful tests using the Clark hot water process in 1948/49 then closed, partly because the recent Leduc discoveries had lessened interest in the oil sands.

Great Canadian Oil Sands

In 1962, Great Canadian Oil Sands Limited (GCOS) received approval from the Alberta government to build and operate a 10,000 cubic metre per day plant near Fort McMurray. The plant was to produce 240 tonnes of sulphur and 900 tonnes of coke per day as by-products. Because at that time the industry was having difficulties marketing its oil, the provincial government established a policy that would limit oil sands production. According to this policy, synthetic oil from the oil sands could supplement conventional oil sales, but could not displace it. Oil from the plant could not exceed 5 per cent of total volumes in markets already supplied by conventional Alberta oil. Financial difficulties delayed construction of the GCOS plant until a new investor - Sun Oil Company's Canadian subsidiary, today known as Suncor - was found. The capacity of the proposed plant increased to 7,500 cubic metres per day and the cost escalated from $122 to $190 million. The larger plant received approval in 1964 and went into commercial production in September 1967. The final cost: $250 million.

During the opening ceremonies for the plant, Sun Oil Company chairman J. Howard Pew (then 85 years old) made remarks which still ring true:

> *"No nation can long be secure in this atomic age unless it be amply supplied with petroleum.... It is the considered opinion of our group that if the North American continent is to produce the oil to meet its requirements in the years ahead, oil from the Athabasca area must of necessity play an important role."*

The Suncor plant was a landmark in oil sands development. It pioneered technology for bitumen extraction and upgrading, and it was the world's first large-scale commercial plant. In the early years it was not particularly profitable, but the plant was nonetheless able to cover operating expenses from the sale of its own production. And in 1979, when federal policy permitted the company to charge world price for its oil, the plant finally became a money-making asset to Suncor. The plant found solutions to the problems of extracting a commercial grade of oil from the sands - problems that had been the concern of financiers, chemists, petroleum engineers, metallurgists, mining engineers, geologists, physicists and many other scientists and pseudo-scientists for many decades.

Syncrude

In 1962 (the same year the Great Canadian Oil Sands proposal went up for approval) Cities Service Athabasca Inc. proposed a 16,000 cubic metre per day plant at the site of its Mildred Lake pilot project. Including a pipeline to Edmonton, the plant was to cost $56 million, with construction beginning in 1965 and completion in 1968. However, the Oil and Gas Conservation Board had concerns about competition between synthetic oil and conventional oil for limited markets. It therefore decided not to bring too many oil sands plants on stream at once, and rejected the Cities Service proposal in favour of the GCOS project.

Cities Service later reapplied for a much larger plant, and the proposal received approval in late 1969. The Syncrude plant which resulted went on production exactly two centuries after Peter Pond's first sighting of the oil sands, in 1978. But before the plant shipped its first barrel of oil, the project went through many trials.

The reason for the long gap between approval and completion was an alarming escalation of costs that beset all major North American projects in the 1970s. High inflation multiplied budgets for practically every aspect of the Syncrude project.

Reviewing project costs in late 1973, the Syncrude consortium found that costs had more than doubled, from $1 billion to $2.3 billion. In December 1974, Atlantic Richfield (whose American parent needed cash to develop its Prudhoe Bay interests) withdrew its 30 per cent participation in the project. A few days later, the three remaining partners informed the Alberta government that the maximum risk they were willing to take on the project was $1 billion. They would need to find another $1 billion of risk capital if the project were to go on. The alternative - shutting down the project - would have cost the four partners (Including Atlantic Richfield) an estimated $250 million.

By this time the world was in the thralls of an energy crisis. Beginning in 1973, the members of the Organisation of Petroleum Exporting Countries had taken advantage of tight world oil supplies to rapidly and regularly increase prices. Policy-makers in the oil consuming countries therefore considered it a matter of national urgency to develop stable, secure energy supplies. Because the resource was so large and development was clearly possible, the oil sands looked like Canada's best bet. As a result, the prospect that the Syncrude project would collapse was a matter of both political and economic concern.

An executive group representing the remaining partners invited the other governments of Canada to participate as commercial partners in the project.

The province also reviewed the cost estimate given by the oil companies. When it found that the consortium's cost estimates were not out of line, the governments of Canada, Alberta and Ontario participated in a historic meeting in Winnipeg in February, 1975. That meeting salvaged the project.

The federal government took a 15 per cent interest, Alberta 10 per cent and Ontario five per cent. The private partners - Cities Service Canada, Gulf Oil Canada and Imperial Oil - agreed to retain their $1.4 billion interest in the project, but gave Alberta the option to convert a $200 million loan to Gulf and Cities Service into ownership interests. Alberta also took full ownership in the no-risk pipeline and electrical utility which the plant needed.

The plant went into operation in the summer of 1978 and produced 5 million barrels (790,000 m^3) of oil within a year. World oil prices leaped skyward in 1979-80 and remained high for the first half of the 1980s. This helped Syncrude become successful financially as well as technically. Syncrude now meets about 14 per cent of Canada's oil requirements, mostly in the form of synthetic oil. The plant has produced nearly 2 billion barrels (320,000,000 m^3) of this oil.

Shell in the Oil Sands

In 2003, Shell Canada and its partners began producing from the Muskeg River Mine, located 75 kilometres north of Fort McMurray. Known as the Athabasca Oil Sands Project, the entire complex consists of Muskeg River, Shell's Scotford Upgrader located near Fort Saskatchewan, Alberta, and supporting facilities.

Four years later, by which time Shell Canada had been wholly acquired by its parent, Royal Dutch Shell, the company applied to build a massive oil sands upgrading complex at the site of its Edmonton refinery. The project, which could cost as much as $27-billion, would be built in four 100,000-barrel-per-day (16,000 m^3/d) stages. Like its existing upgrader, Shell's new facility would process bitumen from the Athabasca Oil Sands Project, as well as bitumen from steam-driven *in situ* oil sands projects.

In Situ Recovery

The oilsands projects just described are unique in the world: They exploit near-surface bitumen from open-pit mines. The industry has also spent decades experimenting with ways to recover bitumen from deeper deposits. The only way to develop petroleum resources underground is through *in situ* production techniques.

In situ means "in place," and refers to recovery techniques which apply heat or solvents to oil reservoirs beneath the earth. There are several varieties of *in situ* technique, but the ones that work best in the oil sands use heat.

The first *in situ* experiment in Alberta took place in 1910, when a Pittsburgh-based outfit, the Barber Asphalt and Paving Company, drilled a bore hole into the bitumen and pumped in steam to liquefy the oil. The experiment failed. In the early 1920s, other *in situ* experiments also took place, but none were commercially successful.

Jacob Owen Absher: In the mid-1920s, a remarkable and persistent experimenter named Jacob Owen Absher incorporated the Bituminous Sand Extraction Company. In 1926, Absher received a Canadian patent for his *in situ* experiments, and he carried on numerous experiments over the following five years - efforts that drew the interest of oil sands pioneers Sidney Ells and Karl Clark. Absher not only used steam to melt the bitumen, but also tried igniting fires within his wells. In the end, however, he was unable to produce oil from the oil sands. His activities ended as the Great Depression raged.

While Absher has been largely forgotten as a pioneer in the oil sands business, others have realised his dream of using heat to release oil from the sands. Today, some commercial projects pipe high-pressure steam into the oil sands reservoir. Other projects actually ignite the oil underground, then pump air below the surface to keep combustion going. These techniques effectively melt the oil, which pumps then bring to the surface.

Thermonuclear thinking: The most dramatic proposal for *in situ* production from deep oil sand deposits came from Richfield Oil Company. In 1959 Richfield suggested an experimental plan to release liquid hydrocarbons from the sand through the expedient of an underground nuclear explosion. The company proposed detonating a 9-kiloton explosive device below the oil sands at a site 100 kilometres south of Fort McMurray. Thermonuclear heat would create a large underground cavern and simultaneously liquefy the oil. The cavern could serve as a collection point for the now-fluid oil, enabling the company to produce it. This idea came remarkably close to reality. The project received federal approval in Canada, and the United States Atomic Energy Commission agreed to provide the device. But before the experiment could take place, public pressure for an international ban on nuclear testing had mounted. The provincial government withheld approval and thus killed the plan.

In situ bitumen production: Many companies experimented with thermal techniques to produce heavy oil from the oilsands, especially in the Cold Lake oilsands deposit, in the 1970s and 1980s. Bearing such field-hand monikers as "steam flood", "fire flood" and "huff and puff" techniques, these extraction methods - like the Barber Asphalt and Paving Company's 1910 experiment - essentially apply heat to the underground reservoir. This melts the oil - that is, decreases its viscosity - so it can be pumped to the surface. An increasingly successful system now in use is steam assisted gravity drainage (SAGD).

SAGD was initially tested at the Underground Test Facility (UTF), ah experimental bitumen mining project funded by AOSTRA and officially opened on June 29th, 1987. The magnitude of the UTF is hard to imagine. Sinking the shafts was done with a drill bit almost four metres in diametre, weighing 230 tonnes. The two shafts below the oil sand reservoirs were 223 metres deep and neither one deviated from the vertical by more than 25 mm. As a safety measure, AOSTRA constructed two parallel tunnels through the limestone under the oil sand reservoir. More than a kilometre in length, each tunnel was five metres wide by four metres high.

From the tunnels the researchers drilled wells up into the reservoir to conduct two sets of tests. The Phase A pilot involved three well pairs 70 metres in length, each with 40-50 metres of exposure to the McMurray formation. Phase B involved another three well pairs, 70 metres apart, each with 500 to 550 metres of direct contact with the oil sand reservoir. The results were excellent, and the petroleum industry soon began producing bitumen through SAGD well pairs drilled and operated from the surface.

The largest single plant in Canada to use *in situ* production is Imperial Oil's Cold Lake oil sand plant. This plant uses a technique called cyclic steam injection. Using this method, the company pumps high-pressure steam into a section of the underground reservoir for a week or so, then pumps the liquid oil out for as long as several months. Imperial also uses steam-assisted gravity drainage. In its SAGD production system, Imperial drills two horizontal wells, one five metres above the other. Steam injected through the upper well reduces the viscosity of the oil, which is recovered through the lower borehole. This plant produces more than 150,000 barrels (24,000 m^3) of bitumen per day.

The first Asian-owned company involved in the oil sands was JACOS, which in 1978 began participating in experiments at a pilot project in the Athabasca area. Like Imperial at Cold Lake, from 1984 to 1994 JACOS and its partners also experimented with a cyclic steam stimulation pilot project on the Hangingstone Lease. Since then the company has developed SAGD production on that lease. It is also constructing a demonstration plant using solvent-based *in situ* bitumen extraction.

Heavy Oil

Heavy oil is a sister resource to bitumen. It is lighter than bitumen and its reservoirs are much smaller than the great oil sands deposits. Like the oil sands, only a small percentage of Canada's large heavy oil resource is producible.

Often called conventional heavy oil, this low-density oil can be recovered by conventional drilling techniques or by waterflood, a technique of injecting water into the reservoir to increase pressure, thus forcing the oil toward the well bore. When these techniques work, heavy oil is like the more commercially attractive lighter grades of oil. But heavy oil can also be quite viscous. It can need some form of heat or solvent and pressure before it can flow into a well bore to be produced. When heavy oil requires these techniques to go into production, it is known as non-conventional heavy oil.

The first heavy oil discoveries came with the pursuit of conventional light and medium crude oil. Because much of western Canada's heavy oil is in pools close to the surface, early explorers using older rigs discovered many of those pools before they came upon the deeper light oil reservoirs.

One of the first finds was in the Ribstone area near Wainwright, Alberta in 1914. The province's first significant production of heavy oil came from the Wainwright field in 1926. Producers drew almost 6,000 barrels (950 m^3) of heavy oil from the field in that year. A small-scale local refinery distilled the heavy goo into usable products. Elsewhere in Alberta, petroleum explorers made other heavy oil finds as they pursued the elusive successor to the Turner Valley oil field. They developed production from many of these fields, but only in small volumes. The recovery techniques of the day combined with the low price of oil and the nature and size of the finds meant that most of the oil remained undeveloped.

Husky

The most important exception was at Lloydminster. While the first discovery occurred in 1938, serious development did not begin until Husky Oil moved into the area after the second world war.

Husky Oil was born during the Depression through the efforts of Glenn Nielson, an Alberta farmer driven to bankruptcy when the bank called a loan on his farm. Nielson had moved to Cody, Wyoming, by the time he founded Husky as a refining operation. He turned his attention back to Canada after the second world war, and decided to set up a refinery at Lloydminster. Steel was scarce, so Husky dismantled a small

Wyoming refinery constructed during the war to provide bunker fuel to the American Navy. It loaded the pieces onto 40 gondola cars and shipped them north by railway.

The company began reassembling the 400 cubic metre per day facility in 1946, and the refinery went on production the following year. Strategically located between the Canadian Pacific and Canadian National railroad tracks in Lloydminster, the refinery soon began to get contracts for locomotive bunker fuel. The company also found a strong market for asphalt for road building.

Husky's move into the area spurred drilling and production. Within two years of Husky's arrival, there were oversupplies of heavy oil and shortages of storage space. Producers solved the problem by storing the oil in earthen pits holding up to 16,000 cubic metres each. For a while Husky bought the oil by weight rather than volume since it was clogged with earth, tumble weed and jackrabbits. The company had to strain and remeasure the stuff before it could begin refining.

Husky began producing heavy oil from local fields in 1946, and by the 1960s was easily the biggest regional producer. In 1963 the company undertook another in a series of expansions to the refinery. To take advantage of expanding markets for Canadian oil, it also began a program to deliver heavy oil to national and export markets.

The key to the $35 million project was the construction of a reversible pipeline which could move the viscous heavy oil into the marketplace. The 116-kilometre "yo-yo" pipeline - the first in the world - brought condensate from the Interprovincial Pipe Line station at Hardisty, Alberta. The company began mixing this very light hydrocarbon with heavy oil, enabling it to flow more easily. The company then pumped the blend through its pipeline (hence the nickname "yo-yo") back to Hardisty. From there the Interprovincial took it eastward to market.

These developments made heavy oil for the first time more than a marginal resource. Within five years, area production had increased fivefold to nearly 2,000 cubic metres per day. By the early 1990s, production from the heavy oil belt was some 40,000 cubic metres per day, and Husky was still one of Canada's biggest heavy oil producers.

Upgraders

Heavy crude feedstock needs pre processing before it is fit for conventional refineries. This is called 'upgrading,' the key components of which include 1) removal of water, sand, physical waste and lighter products; 2) catalytic purification (hydrodemetallisation, hydrodesulphurisation and hydrodenitrogenation; and 3) hydrogenation though carbon rejection or catalytic hydrocracking. Since carbon rejection is generally inefficient and wasteful, catalytic hydrocracking is preferred in most cases.

Catalytic purification and hydrocracking are together known as hydroprocessing. The big challenge in hydroprocessing is to deal with the impurities found in heavy crude, as they poison the catalysts over time. Many efforts have been made to deal with this to ensure high activity and long life of a catalyst. Catalyst materials and pore size distributions need to be optimised to deal with these challenges.

Figuratively speaking, technological improvements and new infrastructure cause heavy oil reservoirs to grow. Enhanced recovery techniques are urging a higher percentage of the reservoirs' oil to the surface. Research and development are creating technologies which have increased the amount producers can extract. Small improvements in technology applied to such a huge resource could mean enormous additions to Canada's recoverable crude oil reserves.

Few Canadian refineries can process more than small amounts of heavy oil, so production has traditionally gone to United States asphalt plants. This changed in the 1980s, however, with the announcement that construction would begin on two heavy oil upgraders. Like the plants at Syncrude, Suncor and Shell's Scotford facility near Edmonton, these refinery-like operations turn heavy oil and bitumen into lighter and lower-sulphur, more desirable crude.

In the late 1970s, a group of heavy oil producers (Gulf, Husky, Shell, PetroCanada and SaskOil) proposed the Plains Upgrader. This facility would have cost $1.2 billion and upgraded 50,000 barrels (7,900 m^3) of oil per day. Gradually, however, consortium members pulled out of the project as they concluded that the high cost of upgrading would make the project uneconomic. In the end, only PetroCanada and Saskoil - both Crown corporations - remained.

The private sector partners pulled out of the Plains Upgrader because upgrading heavy oil at that time was a risky financial proposition. To be economic, these projects rely on substantial differences in pricing ("differentials") between light and heavy oil. Heavy oil is worth less than light oil; the question is, How much less? Unless upgraded oil fetched considerably more per barrel than the less attractive heavy oil, the upgrader would not make money on processing the stuff.

While the Plains partnership collapsed, the idea survived.

Co-op Upgrader

Their partners gone, SaskOil suggested reducing upgrader costs by integrating with the Consumers' Cooperative Refinery in Regina. This would eliminate duplication in facilities and infrastructure by taking advantage of existing land, processing units, storage and pipeline facilities, technical and operating staff and management.

The Co-op refinery was a product of the Co-operative movement, which began in Britain in the mid-19th century. Frequently expanded and modernised, the Co-op plant (first constructed in 1935) was a small but modern refinery when talk about a refinery/upgrader complex began in the early 1980s. Both the federal and Saskatchewan governments had forbidden their Crown corporations to participate in the project, yet both took part themselves. The province had a particular interest, since an upgrader would increase the market for heavy oil from Saskatchewan's fields. This would give the provincial oil industry an important boost. The federal government saw the project as an opportunity to move the nation one small step towards the stated goal of crude oil self-sufficiency. For its part, the Co-op wanted an assured supply of crude oil for its refinery.

Accordingly, Saskatchewan took a 20 per cent equity position and guaranteed loans equal to 45 per cent of the project. In exchange, it became a 50 per cent partner in the combined operation with Consumers' Co-op, which committed its existing refinery (valued at $500 million) to the project. The federal government guaranteed loans equal to 35 per cent of the project. Repayment on the principal of the loans would not begin until late 1992.

On stream in 1988, the Consumers' Co-op refinery/upgrader complex was a 50,000-barrel-per-day (7,900 m^3/d) facility. The $700 million upgrader provided upgraded oil as refinery feedstock.

Husky Upgrader

The company with the most extensive experience in the heavy oil belt was the one to propose - and eventually develop - Canada's other heavy oil upgrader. Husky began to prepare for the upgrader by building a new 25,000-barrel-per-day (4,000 m^3/d) refinery next to the old plant. This facility - which processed heavy oil into asphalt and simultaneously provided light oils for refining into high-end products like gasoline - was completed in 1983.

After a series of false starts, in 1988 Husky and its three partners announced a firm agreement to construct the Bi-Provincial Upgrader - today better known as the Husky Upgrader. Located just east of

Lloydminster, this $1.6 billion upgrader received most of its funding from government. Originally budgeted at $1.2 billion, the federal, Alberta and Saskatchewan governments owned 31.67 per cent, 24.16 per cent and 17.5 per cent each. The balance belonged to Husky, which has since acquired the entire facility.

Under the terms of the original agreement, Husky would receive 50 per cent of the plant's net revenue plus a 10 per cent return on investment until Husky recovered that investment. The balance of plant profit would go proportionally to Husky's partners. A wrinkle in this arrangement occurred as the project neared completion, however, when Saskatchewan's newly-installed NDP government refused to pay its share of $190 million in cost overruns. The other players eventually agreed to pay Saskatchewan's share, but would withhold returns to that province until they had recovered Saskatchewan's arrears.

The upgrader went on stream in mid-1992, but required debottlenecking before it could reach design capacity of 46,000 barrels per day (7,300 m^3/d). The plant upgrades Lloydminster-area heavy oil and Cold Lakebitumen, making still more of those resources available for central Canadian and American markets.

Heavy oil differentials explain the large cost discrepancies between the Husky Upgrader ($1.6 billion for 46,000 barrels per day (7,300 m^3/d) capacity) and the Co-op upgrader ($600 million for 50,000 barrels per day (7,900 m^3/d).) The Husky facility was designed to process heavier grades of oil than the Co-op upgrader, and its output was more desirable. This critical difference meant Husky would pay less for its feedstock and receive more for its output than the Co-op plant. From the beginning, forecasts about these differentials were vital factors in economic calculations for the two projects, each of which has since undergone major expansions.

Metric Conversions

One cubic metre of oil = 6.29 barrels. One cubic metre of natural gas = 35.49 cubic feet. One kilopascal = 1% of atmospheric pressure (near sea level).

Canada's oil measure, the cubic metre, is unique in the world. It is metric in the sense that it uses metres, but it is based on volume so that Canadian units can be easily converted into barrels. In the rest of the metric world, the standard for measuring oil is the metric tonne. The advantage of the latter measure is that it reflects oil quality. In general, lower grade oils are heavier.

Oil Megaprojects

Summary of Megaprojects

A series of project tabulations and analyses by Chris Skrebowski, or of *Petroleum Review*, have presented a more pessimistic picture of future oil supply. In a 2004 report, based on an analysis of new projects over 100,000 barrels per day (16,000 m^3/d), he argued that although ample supply might be available in the near-term, after 2007 "the volumes of new production for this period are well below likely requirements." By 2006, although "the outlook for future supply appears somewhat brighter than even six months ago", nonetheless, if "all the factors reducing new capacity come into play, markets will remain tight and prices high. Only if new capacity flows into the system rather more rapidly than of late, will there be any chance of rebuilding spare capacity and softening prices."

The smallest fields, even in aggregate, do not contribute a large fraction of the total. For example, a relatively small number of giant and super-giant oilfields are providing almost half of the world production.

Megaprojects predicted for individual years;

Over view	20 03	20 04	20 05	20 06	20 07	20 08	20 09	20 10	20 11	20 12	20 13	20 14	20 15	20 16	20 17	20 18	20 19	20 20

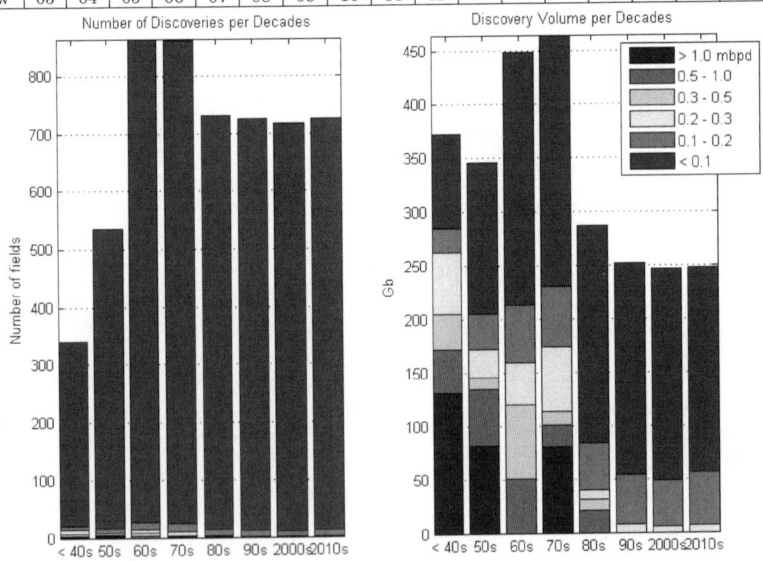

Figure : *Number of oil fields discovered per decades grouped by average flow rates (left) and corresponding oil volumes (right) in giga-barrels (Gb). Data taken from the annexe B of "Twilight in the Desert" by Matthew Simmons.*

Decline Rates

The most important variable is the average decline rate for Fields in Production (FIP) which is difficult to assess.

Giant Oil and Gas Fields

The world's 932 giant oil and gas fields are considered those with 500 million barrels (79,000,000 m^3) of ultimately recoverable oil or gas equivalent. Geoscientists believe these giants account for 40 percent of the world's petroleum reserves. They are clustered in 27 regions of the world, with the largest clusters in the Persian Gulf and Western Siberian Basin. The past three decades reflect declines in discoveries of giant fields. The present decade (2000–2011), however, reflects an upturn in discoveries and appears on track to be the third best for discovery of giant oil and gas fields in the 150 year history of modern oil and gas exploration.

According to analysis led by Paul Mann of the University of Texas' Jackson School of Geosciences, almost all of the 932 giant oil and gas fields cluster within 27 regions, or about 30 percent of Earth's land surface. Since 2003, Mann and colleagues M.K. Horn and Ian Cross have tracked the giants on a map that highlights the tectonic and sedimentary basin maps of the 27 key regions.

Recent work in tracking giant oil and gas fields follows the earlier efforts of the late exploration geologist Michel T. Halbouty, who tracked trends in giant discoveries from the 1960s to 2004.

Tectonic Settings

Geophysicists and exploration geologists who look for oil and gas fields classify the subsurface characteristics, or tectonic setting, of geological structures that contain hydrocarbons. Any one oil and gas field may reflect influences from multiple geological periods and events, but geoscientists often attempt to characterise a field based on the dominant geological event that influenced the structure's ability to trap and contain oil and gas in recoverable quantities.

A majority of the world's giant oil and gas fields exist in two characteristic tectonic settings—passive margin and rift environments. Passive margins are found along the edges of major ocean basins, such as the Atlantic coast of Brazil where oil and gas has been located in large quantities in the Campos basin. Rifts are oceanic ridges formed when tectonic plates separate and a new crust is created. The North Sea is an example of a rift setting associated with prodigious hydrocarbon reserves. Geoscientists theorize that both zones are especially conducive to forming giant oil and gas fields when they are distant from active tectonic areas. Stability appears to be conducive to trapping and retaining hydrocarbons under the subsurface.

Four other common tectonic settings, including collisional margins, strike-slip margins, and subduction margins, are associated with the formation of giant oil and gas fields, though not to the dominant extent of passive margin and rift settings.

Recent and Future Giants

Based on the locations of past giants, Mann et al. predicted new discoveries of giant oil and gas fields would mainly be made in passive margin and rift environments, especially in deepwater basins. They also predicted that existing areas that have produced giant fields would be likely targets for new discoveries of "elephants," as the fields are sometimes known in the oil and gas industry. Data from 2000-2007 reflect the accuracy of their predictions. The 79 new giant oil and gas fields discovered from 2000-2007 tended to be located in similar tectonic settings as the previously documented giants from 1868–2000, with 36 percent along passive margins, 30 percent in rift zones or overlying sags (structures associated with rifts), and 20 percent in collisional zones.

Despite a recent uptick in the number of giant oil and gas fields, discovery of giants appears to have peaked in the 1960s and 1970s. Looking to the future, geoscientists foresee a continuation of the recent trend of discovering more giant gas fields than oil fields. Two major continental regions—Antarctica and the Arctic—remain largely unexplored. Beyond them, however, trends suggest that remaining giant fields will be discovered in "in-fill" areas where past giants have been clustered and in frontier, or new, areas that correspond to the predominant tectonic settings of past giants.

Giant Field Production Properties and Behaviour

Comprehensive analysis of the production from the majority of the world's giant oil fields has shown their enormous importance for global oil production. For instance, the 20 largest oil fields in the world alone account for roughly 25% of the total oil production. The majority of the world's giants are already in decline and this will have dramatic consequences for future oil supply.

Further analysis shows that giant oil fields typically reach their maximum production before 50% of the ultimate recoverable volume has been extracted. A strong correlation between depletion and the rate of decline was also found in that study, indicating that much new technology has only been able to temporarily decrease depletion at the expense of rapid future decline. This is exactly the case in the Cantarell Field.

Gas Compressor

A gas compressor is a mechanical device that increases the pressure of a gas by reducing its volume.

Compressors are similar to pumps: both increase the pressure on a fluid and both can transport the fluid through a pipe. As gases are compressible, the compressor also reduces the volume of a gas. Liquids are relatively incompressible; while some can be compressed, the main action of a pump is to pressurise and transport liquids.

Types of Compressors

The main types of gas compressors are illustrated and discussed below:

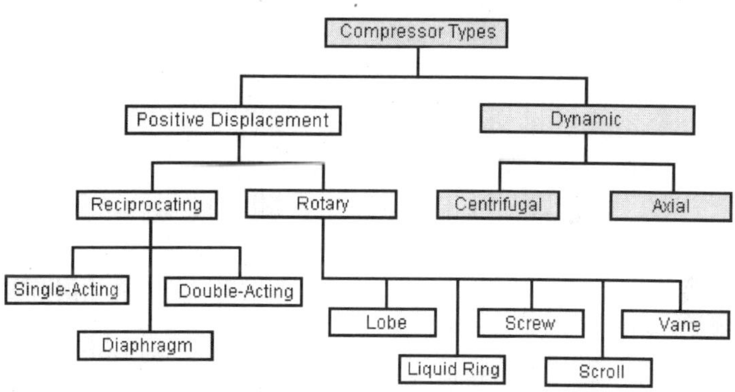

Hermetically Sealed, Open, or Semi-Hermetic

Figure: *A small hermetically sealed compressor in a common consumer refrigerator or freezer; it typically has a rounded steel outer shell that is permanently welded shut, and which seals operating gases inside the system. There is no route for gases to leak, such as around motor shaft seals. On this model, the plastic top section is part of an auto-defrost system which uses motor heat to evapourate the water.*

Compressors are often described as being either open, hermetic, or semi-hermetic, to describe how the compressor and motor drive is situated in relation to the gas or vapour being compressed. The industry name for a hermetic is hermetically sealed compressor, while a semi-is commonly called a semi-hermetic compressor.

In hermetic and most semi-hermetic compressors, the compressor and motor driving the compressor are integrated, and operate within the pressurised gas envelope of the system. The motor is designed to operate and be cooled by the gas or vapour being compressed.

The difference between the hermetic and semi-hermetic, is that the hermetic uses a one-piece welded steel casing that cannot be opened for repair; if the hermetic fails it is simply replaced with an entire new unit. A semi-hermetic uses a large cast metal shell with gasketed covers that can be opened to replace motor and pump components.

The primary advantage of a hermetic and semi-hermetic is that there is no route for the gas to leak out of the system. Open compressors rely on either natural leather or synthetic rubber seals to retain the internal pressure, and these seals require a lubricant such as oil to retain their sealing properties.

An open pressurised system such as an automobile air conditioner can leak its operating gases, if it is not operated frequently enough. Open systems rely on lubricant in the system to splash on pump components and seals. If it is not operated frequently enough, the lubricant on the seals slowly evapourates, and then the seals begin to leak until the system is no longer functional and must be recharged. By comparison, a hermetic system can sit unused for years, and can usually be started up again at any time without requiring maintenance or experiencing any loss of system pressure.

The disadvantage of hermetic compressors is that the motor drive cannot be repaired or maintained, and the entire compressor must be removed if a motor fails. A further disadvantage is that burnt out windings can contaminate whole systems requiring the system to be entirely pumped down and the gas replaced. Typically hermetic compressors are used in low-cost factory-assembled consumer goods where the cost of repair is high compared to the value of the device, and it would be more economical to just purchase a new device.

An advantage of open compressors is that they can be driven by non-electric power sources, such as an internal combustion engine or turbine. However, open compressors that drive refrigeration systems are generally not totally *maintenance free* throughout the life of the system, since some gas leakage will occur over time.

Centrifugal Compressors

Centrifugal compressors use a rotating disk or impeller in a shaped housing to force the gas to the rim of the impeller, increasing the velocity of the gas. A diffuser (divergent duct) section converts the velocity energy to pressure energy. They are primarily used for continuous, stationary service in industries such as oil refineries, chemical and petrochemical plants and natural gas processing plants. Their application can be from 100 horsepower (75 kW) to thousands of horsepower. With multiple staging, they can achieve extremely high output pressures greater than 10,000 psi (69 MPa).

Figure: *A single stage centrifugal compressor*

Many large snow making operations (like ski resorts) use this type of compressor. They are also used in internal combustion engines as superchargers and turbochargers. Centrifugal compressors are used in small gas turbine engines or as the final compression stage of medium sized gas turbines. Sometimes the capacity of the compressors is written in NM3/hr. Here 'N' stands for normal temperature pressure (20°C and 1 atm) for example 5500 NM3/hr.

Diagonal or Mixed-flow Compressors

Diagonal or mixed-flow compressors are similar to centrifugal compressors, but have a radial and axial velocity component at the exit from the rotor. The diffuser is often used to turn diagonal flow to an axial rather than radial direction.

Axial-flow Compressors

Axial-flow compressors are dynamic rotating compressors that use arrays of fan-like airfoils to progressively compress the working fluid. They are used where there is a requirement for a high flow rate or a compact design.

The arrays of airfoils are set in rows, usually as pairs: one rotating and one stationary. The rotating airfoils, also known as blades or *rotors*, accelerate the fluid. The stationary airfoils, also known as *stators* or vanes, decelerate and redirect the flow direction of the fluid, preparing it for the rotor blades of the next stage. Axial compressors are almost always multi-staged, with the cross-sectional area of the gas passage diminishing along the compressor to maintain an optimum axial Mach number. Beyond about 5 stages or a 4:1 design pressure ratio, variable geometry is normally used to improve operation.

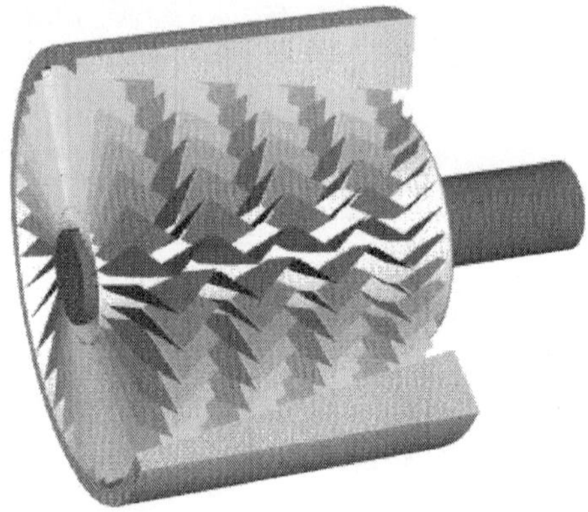

Figure: *An animation of an axial compressor.*

Axial compressors can have high efficiencies; around 90% polytropic at their design conditions. However, they are relatively expensive, requiring a large number of components, tight tolerances and high quality materials. Axial-flow compressors can be found in medium to large gas turbine engines, in natural gas pumping stations, and within certain chemical plants.

Reciprocating Compressors

Figure: *A motor-driven six-cylinder reciprocating compressor that can operate with two, four or six cylinders.*

Reciprocating compressors use pistons driven by a crankshaft. They can be either stationary or portable, can be single or multi-staged, and can be driven by electric motors or internal combustion engines. Small reciprocating compressors from 5 to 30 horsepower (hp) are commonly seen in automotive applications and are typically for intermittent duty. Larger reciprocating compressors well over 1,000 hp (750 kW) are commonly found in large industrial and petroleum applications. Discharge pressures can range from low pressure to very high pressure (>18000 psi or 180 MPa). In certain applications, such as air compression, multi-stage double-acting compressors are said to be the most efficient compressors available, and are typically larger, and more costly than comparable rotary units. Another type of reciprocating compressor is the swash plate compressor, which uses pistons which are moved by a swash plate mounted on a shaft.

Household, home workshop, and smaller job site compressors are typically reciprocating compressors 1½ hp or less with an attached receiver tank.

Rotary Screw Compressors

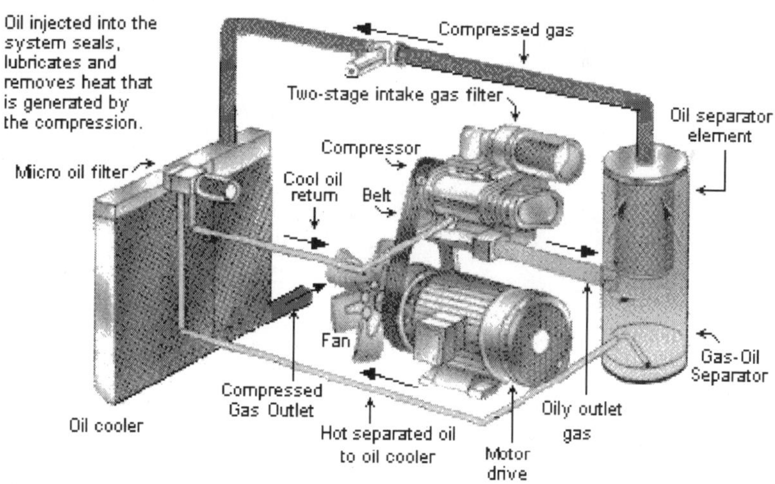

Figure: *Diagram of a rotary screw compressor*

Rotary screw compressors use two meshed rotating positive-displacement helical screws to force the gas into a smaller space. These are usually used for continuous operation in commercial and industrial applications and may be either stationary or portable. Their application can be from 3 horsepower (2.2 kW) to over 1,200 horsepower (890 kW) and from low pressure to moderately high pressure (>1,200 psi or 8.3 MPa). Rotary screw compressors are commercially produced in Oil Flooded, Water Flooded and Dry type.

Rotary Vane Compressors

Rotary vane compressors consist of a rotor with a number of blades inserted in radial slots in the rotor. The rotor is mounted offset in a larger housing which can be circular or a more complex shape. As the rotor turns, blades slide in and out of the slots keeping contact with the outer wall of the housing. Thus, a series of decreasing volumes is created by the rotating blades. Rotary Vane compressors are, with piston compressors one of the oldest of compressor technologies.

With suitable port connections, the devices may be either a compressor or a vacuum pump. They can be either stationary or portable, can be single or multi-staged, and can be driven by electric motors or internal combustion engines. Dry vane machines are used at relatively low pressures (e.g., 2 bar or 200 kPa; 29 psi) for bulk material movement while oil-injected machines have the necessary volumetric efficiency to achieve pressures up to about 13 bar (1,300 kPa; 190 psi) in a single stage. A rotary vane compressor is well suited to electric motor drive and is significantly quieter in operation than the equivalent piston compressor.

Rotary vane compressors can have mechanical efficiencies of about 90%.

Scroll Compressors

A scroll compressor, also known as scroll pump and scroll vacuum pump, uses two interleaved spiral-like vanes to pump or compress fluids such as liquids and gases. The vane geometry may be involute, archimedean spiral, or hybrid curves. They operate more smoothly, quietly, and reliably than other types of compressors in the lower volume range.

Often, one of the scrolls is fixed, while the other orbits eccentrically without rotating, thereby trapping and pumping or compressing pockets of fluid or gas between the scrolls.

This type of compressor was used as the supercharger on Volkswagen G60 and G40 engines in the early 1990s.

Diaphragm Compressors

A diaphragm compressor (also known as a membrane compressor) is a variant of the conventional reciprocating compressor. The compression of gas occurs by the movement of a flexible membrane, instead of an intake element. The back and forth movement of the membrane is driven by a rod and a crankshaft mechanism. Only the membrane and the compressor box come in contact with the gas being compressed.

The degree of flexing and the material constituting the diaphragm affects the maintenance life of the equipment. Generally stiff metal diaphragms may only displace a few cubic centimetres of volume because the metal can not endure large degrees of flexing without cracking, but the stiffness of a metal diaphragm allows it to pump at high pressures. Rubber or silicone diaphragms are capable of enduring deep pumping strokes of very high flexion, but their low strength limits their use to low-pressure applications, and they need to be replaced as plastic embrittlement occurs.

Diaphragm compressors are used for hydrogen and compressed natural gas (CNG) as well as in a number of other applications.

Figure: *A three-stage diaphragm compressor*

The photograph included in this section depicts a three-stage diaphragm compressor used to compress hydrogen gas to 6,000 psi (41 MPa) for use in a prototypecompressed hydrogen and compressed natural gas (CNG) fuelling station built in downtown Phoenix, Arizona by the Arizona Public Service company (an electric utilities company). Reciprocating compressors were used to compress the natural gas. The prototype alternative fuelling station was built in compliance with all of the prevailing safety, environmental and building codes in Phoenix to demonstrate that such fuelling stations could be built in urban areas.

Air Bubble Compressor

Also known as a trompe. A mixture of air and water generated through turbulence is allowed to fall into a subterranean chamber where the air separates from the water. The weight of falling water compresses the air in the top of the chamber. A submerged outlet from the chamber allows water to flow to the surface at a lower height than the intake. An outlet in the roof of the chamber supplies the compressed air to the surface. A facility on this principal was built on the Montreal River at Ragged Shutes near Cobalt, Ontario in 1910 and supplied 5,000 horsepower to nearby mines.

Temperature

Gas Laws: Compression of a gas naturally increases its temperature, often referred to as the *heat of compression.*

$$W = \int_{V_1}^{V_2} PdV = P_1V_1^n \int_{V_1}^{V_2} V^{-n}dV$$

where

$$\frac{P_2}{P_1} = \left(\frac{v_1}{v_2}\right)^n$$

so

$$W = \frac{P_1V_1^n}{1-n}(V_2^{1-n} - V_1^{1-n})$$

with n taking different values for different compression processes.

- Adiabatic - This model assumes that no energy (heat) is transferred to or from the gas during the compression, and all supplied work is added to the internal energy of the gas, resulting in increases of temperature and pressure. Theoretical temperature rise is:

$$T_2 = T_1.R_c^{(k-1)/k}$$

with T_1 and T_2 in degrees Rankine or kelvins, and k = ratio of specific heats (approximately 1.4 for air). R_c is the compression ratio; being the absolute outlet pressure divided by the absolute inlet pressure. The rise in air and temperature ratio means compression does not follow a simple pressure to volume ratio. This is less efficient, but quick. Adiabatic compression or expansion more closely model real life when a compressor has good insulation, a large gas volume, or a short time scale (i.e., a high power level). In practice there will always be a certain amount of heat flow out of the compressed gas. Thus, making a perfect adiabatic compressor would require perfect heat insulation of all parts of the machine. For example, even a bicycle tire pump's metal tube becomes hot as you compress the air to fill a tire. The relation between temperature and compression ratio described above means that the value of n for an adiabatic process is k (the ratio of specific heats).

Isothermal - This model assumes that the compressed gas remains at a constant temperature throughout the compression or expansion process. In this cycle, internal energy is removed from the system as heat at the same rate that it is added by the mechanical work of compression. Isothermal compression or expansion more closely models real life when the compressor has a large heat exchanging surface, a small gas volume, or a long time scale (i.e., a small power level). Compressors that utilise inter-stage cooling between compression stages come closest to achieving perfect isothermal compression. However, with practical devices perfect isothermal compression is not attainable. For example, unless you have an infinite number of compression stages with corresponding intercoolers, you will never achieve perfect isothermal compression.

For an isothermal process, n is 1, so the value of the work integral for an isothermal process is:

$$W = -P_1 v_1 \ln\left(\frac{P_2}{P_1}\right)$$

When evaluated, the isothermal work is found to be lower than the adiabatic work.

Polytropic - This model takes into account both a rise in temperature in the gas as well as some loss of energy (heat) to the compressor's components. This assumes that heat may enter or leave the system, and that input shaft work can appear as both increased pressure (usually useful work) and increased temperature above adiabatic (usually losses due to cycle efficiency). Compression efficiency is then the ratio of temperature rise at theoretical 100 percent (adiabatic) vs. actual (polytropic). Polytropic compression will use a value of n between 0 (a constant-pressure process) and infinity (a constant volume process). For the typical case where an effort is made to cool the gas compressed by an approximately adiabatic process, the value of n will be between 1 and k.

Staged Compression

In the case of centrifugal compressors, commercial designs currently do not exceed a compression ratio of more than a 3.5 to 1 in any one stage (for a typical gas). Since compression generates heat, the compressed gas is to be cooled between stages making the compression less adiabatic and more isothermal. The inter-stage coolers typically result in some partial condensation that is removed in vapour-liquid separators.

In the case of small reciprocating compressors, the compressor flywheel may drive a cooling fan that directs ambient air across the intercooler of a two or more stage compressor.

Because rotary screw compressors can make use of cooling lubricant to remove the heat of compression, they very often exceed a 9 to 1 compression ratio. For instance, in a typical diving compressor the air is compressed in three stages. If each stage has a compression ratio of 7 to 1, the compressor can output 343 times atmospheric pressure ($7 \times 7 \times 7 = 343$ atmospheres). (343 atm/34.8 MPa; 5.04 ksi)

Prime Movers

There are many options for the "prime mover" or motor which powers the compressor:

- gas turbines power the axial and centrifugal flow compressors that are part of jet engines
- steam turbines or water turbines are possible for large compressors
- electric motors are cheap and quiet for static compressors. Small motors suitable for domestic electrical supplies use single phase alternating current. Larger motors can only be used where an industrial electricalthree phase alternating current supply is available.
- diesel engines or petrol engines are suitable for portable compressors and support compressors. Common in automobiles and other types of vehicles (including piston-powered aeroplanes, boats, trucks, etc.), diesel or gasoline engines can power compressors using their own crankshaft power (this setup known as a supercharger), or, using their waste exhaust gas to spin a turbine connected to the compressor (this setup known as a turbocharger).

Applications

Gas compressors are used in various applications where either higher pressures or lower volumes of gas are needed:

- in pipeline transport of purified natural gas to move the gas from the production site to the consumer. Often, the compressor in this application is driven by a gas turbine which is fuelled by gas bled from the pipeline. Thus, no external power source is necessary.

- in petroleum refineries, natural gas processing plants, petrochemical and chemical plants, and similar large industrial plants for compressing intermediate and end product gases.
- in refrigeration and air conditioner equipment to move heat from one place to another in refrigerant cycles.
- in gas turbine systems to compress the intake combustion air
- in storing purified or manufactured gases in a small volume, high pressure cylinders for medical, welding and other uses.
- in many various industrial, manufacturing and building processes to power all types of pneumatic tools.
- as a medium for transferring energy, such as to power pneumatic equipment.
- in pressurised aircraft to provide a breathable atmosphere of higher than ambient pressure.
- in some types of jet engines (such as turbojets and turbofans) to provide the air required for combustion of the engine fuel. The power to drive the combustion air compressor comes from the jet's own turbines.
- in SCUBA diving, hyperbaric oxygen therapy and other life support devices to store breathing gas in a small volume such as in diving cylinders.
- in surface supplied diving an air compressor is frequently used to supply low pressure air (10 to 20 bar) to the diver for breathing
- in submarines, to store air for later use in displacing water from buoyancy chambers, for adjustment of depth.
- in turbochargers and superchargers to increase the performance of internal combustion engines by increasing mass flow.
- in rail and heavy road transport to provide compressed air for operation of rail vehicle brakes or road vehicle brakes and various other systems (doors, windscreen wipers, engine/ gearbox control, etc.).
- in miscellaneous uses such as providing compressed air for filling pneumatic tires.
- in the case of the fire piston and the heat pump, the desired outcome is the temperature rise of the gas, and compressing the gas is only a means to that end.

Natural Gas

Natural gas is a naturally occurring hydrocarbon gas mixture consisting primarily of methane, with up to 20 percent concentration of other hydrocarbons (usually ethane) as well as small amounts of impurities such as carbon dioxide.

Natural gas is widely used and is an important energy source in many applications including heating buildings, generating electricity, providing heat and power to industry and vehicles and is also a feedstock in the manufacture of products such as fertilizers.

Natural gas is found in deep underground natural rock formations or associated with other hydrocarbon reservoirs, in coal beds, and as methane clathrates. Most natural gas was created over time by two mechanisms: biogenic and thermogenic. Biogenic gas is created by methanogenic organisms in marshes, bogs, landfills, and shallow sediments. Deeper in the earth, at greater temperature and pressure, thermogenic gas is created from buried organic material.

Before natural gas can be used as a fuel, it must undergo processing to clean the gas and remove impurities including water in order to meet the specifications of marketable natural gas. The by-products of processing include ethane, propane, butanes, pentanes, and higher molecular weight hydrocarbons, hydrogen sulphide (which may be converted into pure sulfur), carbon dioxide, water vapour, and sometimes helium and nitrogen.

Natural gas is often informally referred to simply as gas, especially when compared to other energy sources such as oil or coal.

Sources

List of natural gas fields, List of countries by natural gas proven reserves, and List of countries by natural gas production

Natural Gas

In the 19th century, natural gas was usually obtained as a byproduct of producing oil, since the small, light gas carbon chains came out of solution as the extracted fluids underwent pressure reduction from the reservoir to the surface, similar to uncapping a bottle of soda pop where the carbon dioxide effervesces. Unwanted natural gas was a disposal problem in the active oil fields. If there was not a market for natural gas near the wellhead it was virtually valueless since it had to be piped to the end user. In the 19th century and early 20th century, such unwanted gas was usually burned off in the oil fields. Today, unwanted gas (or stranded gas without a market) associated with oil extraction often is returned to the reservoir with 'injection' wells while awaiting a possible future market or to repressurise the formation, which can enhance extraction rates from other wells. In regions with a high natural gas demand (such as the US), pipelines are constructed when economically feasible to move the gas from the wellsite to the end consumer.

Another possibility is to export the natural gas as a liquid. Gas-to-liquids (GTL) is a developing technology that converts stranded natural gas into synthetic gasoline, diesel, or jet fuel through the Fischer-Tropsch process developed during World War II by Germany. Such fuel can be transported to users through conventional pipelines and tankers. Proponents claim GTL burns cleaner than comparable petroleum fuels. Most major international oil companies are in an advanced stage of GTL production, with a world-scale (140,000 barrels (22,000 m^3) a day) GTL plant in Qatar scheduled to be in production before 2010.

Natural gas can be "associated" (found in oil fields) or "non-associated" (isolated in natural gas fields), and is also found in coal beds (as coalbed methane). It sometimes contains significant amounts of ethane, propane, butane, and pentane—heavier hydrocarbons removed for commercial use prior to the methane being sold as a consumer fuel or chemical plant feedstock. Non-hydrocarbons such as carbon dioxide, nitrogen, helium (rarely), and hydrogen sulfide must also be removed before the natural gas can be transported. Natural gas is commercially extracted from oil fields and natural gas fields. Gas extracted from oil wells is called casinghead gas or associated gas. The natural gas industry is extracting gas from increasingly more challenging resource types: sour gas, tight gas, shale gas, and coalbed methane.

The world's largest proven gas reserves are located in Russia, with 4.757×10 m^3 (1.68×10cubic feet). With the Gazprom company, Russia is frequently the world's largest natural gas extractor. Major proven resources (in billion cubic metres) are world 175,400 (2006), Russia 47,570 (2006), Iran 26,370 (2006), Qatar 25,790 (2007), Saudi Arabia 6,568 (2006) and United Arab Emirates 5,823 (2006).

It is estimated that there are about 900 trillion cubic metres of "unconventional" gas such as shale gas, of which 180 trillion may be recoverable. In turn, many studies from MIT, Black & Veatch and the DOE will account for a larger portion of electricity generation and heat in the future.

The world's largest gas field is Qatar's offshore North Field, estimated to have 25 trillion cubic metres (9.0×10 cubic feet) of gas in place—enough to last more than 420 years at optimum extraction levels. The second largest natural gas field is the South Pars Gas Field in Iranian waters in the Persian Gulf. Located next to Qatar's North Field, it has an estimated reserve of 8 to 14 trillion cubic metres (2.8×10 to 5.0×10 cubic feet) of gas.

Because natural gas is not a pure product, as the reservoir pressure drops when non-associated gas is extracted from a field under supercritical (pressure/temperature) conditions, the higher molecular weight components may partially condense upon isothermic depressurising—an effect called retrograde condensation. The liquid thus formed may get trapped as the pores of the gas reservoir get deposited. One method to deal with this problem is to re-inject dried gas free of condensate to maintain the underground pressure and to allow re-evapouration and extraction of condensates. More frequently, the liquid condenses at the surface, and one of the tasks of the gas plant to collect this condensate. The resulting liquid is called natural gas liquid (NGL) and has commercial value.

Town Gas

Town gas, a synthetically produced mixture of methane and other gases, mainly the highly toxic carbon monoxide, is used in a similar way to natural gas and can be produced by treating coal chemically. This is a historical technology, not usually economically competitive with other sources of fuel gas today. But there are still some specific cases where it is the best option and it may be so into the future.

Most town "gashouses" located in the eastern US in the late 19th and early 20th centuries were simple by-product coke ovens which heated bituminous coal in air-tight chambers. The gas driven off from the coal was collected and distributed through networks of pipes to residences and other buildings where it was used for cooking and lighting. (Gas heating did not come into widespread use until the last half of the 20th century.) The coal tar (or asphalt) that collected in the bottoms of the gashouse ovens was often used for roofing and other water-proofing purposes, and when mixed with sand and gravel was used for paving streets.

Biogas

When methane-rich gases are produced by the anaerobic decay of non-fossil organic matter (biomass), these are referred to as biogas (or natural biogas). Sources of biogas include swamps, marshes, and landfills, as well as sewage sludge and manure by way of anaerobic digesters, in addition to enteric fermentation, particularly in cattle.

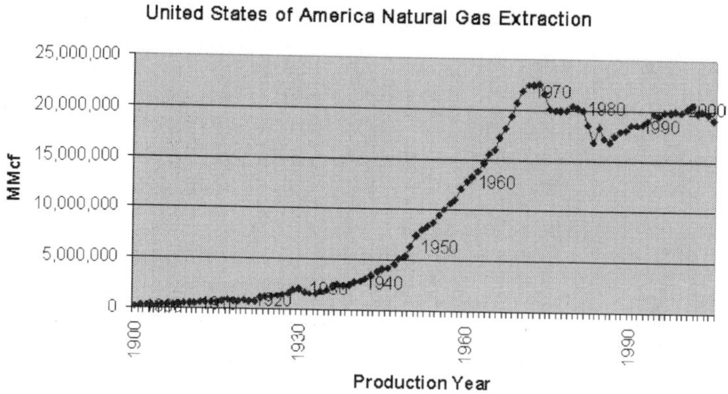

Figure: *U.S. natural gas extraction, 1900–2005. Source: EIA.*

Methanogenic archaea are responsible for all biological sources of methane, some in symbiotic relationships with other life forms, including termites, ruminants, and cultivated crops. Methane released directly into the atmosphere would be considered a pollutant. However, methane in the atmosphere is oxidised, producing carbon dioxide and water. Methane in the atmosphere has a half life of seven years, meaning that if a tonne of methane were emitted today, 500 kilograms would have broken down to carbon dioxide and water after seven years.

Other sources of methane, the principal component of natural gas, include landfill gas, biogas, and methane hydrate. Biogas, and especially landfill gas, are already used in some areas, but their use could be greatly expanded. Landfill gas is a type of biogas, but biogas usually refers to gas produced from organic material that has not been mixed with other waste.

Landfill gas is created from the decomposition of waste in landfills. If the gas is not removed, the pressure may get so high that it works its way to the surface, causing damage to the landfill structure, unpleasant odour, vegetation die-off, and an explosion hazard. The gas can be vented to the atmosphere, flared or burned to produce electricity or heat. Experimental systems were being proposed for use in parts of Hertfordshire, UK, and Lyon in France.

Once water vapour is removed, about half of landfill gas is methane. Almost all of the rest is carbon dioxide, but there are also small amounts of nitrogen, oxygen, and hydrogen. There are usually trace amounts of hydrogen sulfide and siloxanes, but their concentration varies widely. Landfill gas cannot be distributed through utility natural gas pipelines unless it is cleaned up to less than 3% CO_2, and a few parts per million H_2S, because CO_2 and H_2Scorrode the pipelines. It is usually more economical to combust the gas on site or within a short distance of the landfill using a dedicated pipeline. Water vapour is often removed, even if the gas is combusted on site. If low temperatures condense water out of the gas, siloxanes can be lowered as well because they tend to condense out with the water vapour. Other non-methane components may also be removed in order to meet emission standards, to prevent fouling of the equipment or for environmental considerations. Co-firing landfill gas with natural gas improves combustion, which lowers emissions.

Gas generated in sewage treatment plants is commonly used to generate electricity. For example, the Hyperion sewage plant in Los Angeles burns 8 million cubic feet (230,000 m^3) of gas per day to generate power New York City utilises gas to run equipment in the sewage plants, to generate electricity, and in boilers. Using sewage gas to make electricity is not limited to large cities. The city of Bakersfield, California, uses cogeneration at its sewer plants. California has 242 sewage wastewater treatment plants, 74 of which have installed anaerobic digesters. The total biopower generation from the 74 plants is about 66 MW.

Figure: *The McMahon natural gas processing plant in Taylor, British Columbia, Canada.*

Biogas is usually produced using agricultural waste materials, such as otherwise unusable parts of plants and manure. Biogas can also be produced by separating organic materials from waste that otherwise goes to landfills. This method is more efficient than just capturing the landfill gas it produces. Using materials that would otherwise generate no income, or even cost money to get rid of, improves the profitability and energy balance of biogas production.

Anaerobic lagoons produce biogas from manure, while biogas reactors can be used for manure or plant parts. Like landfill gas, biogas is mostly methane and carbon dioxide, with small amounts of nitrogen, oxygen and hydrogen. However, with the exception of pesticides, there are usually lower levels of contaminants.

Crystallised Natural Gas — Hydrates

Huge quantities of natural gas (primarily methane) exist in the form of hydrates under sediment on offshore continental shelves and on land in arctic regions that experience permafrost, such as those in Siberia. Hydrates require a combination of high pressure and low temperature to form. However, as of 2010 no technology has been developed yet to extract natural gas economically from hydrates.

In 2010, using current technology, the cost of extracting natural gas from crystallised natural gas is estimated to 100–200% the cost of extracting natural gas from conventional sources, and even higher from offshore deposits.

Natural Gas Processing

The block flow diagram also shows how processing of the raw natural gas yields byproduct sulfur, byproduct ethane, and natural gas liquids (NGL) propane, butanes and natural gasoline (denoted as pentanes+).

Power Generation

Natural gas is a major source of electricity generation through the use of gas turbines and steam turbines. Most grid peaking power plants and some off-grid engine-generators use natural gas. Particularly high efficiencies can be achieved through combining gas turbines with a steam turbine in combined cycle mode. Natural gas burns more cleanly than other hydrocarbon fuels, such as oil and coal, and produces less carbon dioxide per unit of energy released. For an equivalent amount of heat, burning natural gas produces about 30% less carbon dioxide than burning petroleum and about 45% less than burning coal. Combined cycle power generation using natural gas is thus the cleanest source of power available using hydrocarbon fuels, and this technology is widely used wherever gas can be obtained at a reasonable cost. Fuel cell technology may eventually provide cleaner options for converting natural gas into electricity, but as yet it is not price-competitive.

Domestic Use

Natural gas dispensed from a simple stovetop can generate heat in excess of 2000°F (1093°C) making it a powerful domestic cooking and heating fuel. In much of the developed world it is supplied to homes via pipes where it is used for many purposes including natural gas-powered ranges and ovens, natural gas-heated clothes dryers, heating/cooling, and central heating. Home or other building heating may include boilers, furnaces, and water heaters. Compressed natural gas (CNG) is used in rural homes without connections to piped-in public utility services, or with portable grills. Natural gas is also supplied by independent natural gas suppliers through Natural Gas Choice programs throughout the United States. However, due to CNG being less economical than LPG, LPG (propane) is the dominant source of rural gas.

Transportation

CNG is a cleaner alternative to other automobile fuels such as gasoline (petrol) and diesel. As of 2008 there were 9.6 million natural gas vehicles worldwide, led by Pakistan (2.0 million), Argentina (1.7 million), Brazil (1.6 million), Iran (1.0 million), and India (650,000). The energy efficiency is generally equal to that of gasoline engines, but lower compared with modern diesel engines. Gasoline/petrol vehicles converted to run on natural gas suffer because of the low compression ratio of their engines, resulting in a cropping of delivered power while running on natural gas (10%–15%). CNG-specific engines, however, use a higher compression ratio due to this fuel's higher octane number of 120–130.

Fertilizers

Natural gas is a major feedstock for the production of ammonia, via the Haber process, for use in fertilizer production.

Aviation

Russian aircraft manufacturer Tupolev is currently running a development program to produce LNG- and hydrogen-powered aircraft. The program has been running since the mid-1970s, and seeks to develop LNG and hydrogen variants of the Tu-204 and Tu-334 passenger aircraft, and also the Tu-330 cargo aircraft. It claims that at current market prices, an LNG-powered aircraft would cost 5,000 roubles (~ \$218/ £112) less to operate per ton, roughly equivalent to 60%, with considerable reductions to carbon monoxide, hydrocarbon and nitrogen oxide emissions.

The advantages of liquid methane as a jet engine fuel are that it has more specific energy than the standard kerosene mixes do and that its low temperature can help cool the air which the engine compresses for greater volumetric efficiency, in effect replacing an intercooler. Alternatively, it can be used to lower the temperature of the exhaust.

Hydrogen

Natural gas can be used to produce hydrogen, with one common method being the hydrogen reformer. Hydrogen has many applications: it is a primary feedstock for the chemical industry, a hydrogenating agent, an important commodity for oil refineries, and the fuel source in hydrogen vehicles.

Other

Natural gas is also used in the manufacture of fabrics, glass, steel, plastics, paint, and other products.

Storage and Transport

Because of its low density, it is not easy to store natural gas or transport by vehicle. Natural gas pipelines are impractical across oceans. Many existing pipelines in America are close to reaching their capacity, prompting some politicians representing northern states to speak of potential shortages. In Europe, the gas pipeline network is already dense in the West. New pipelines are planned or under construction in Eastern Europe and between gas fields in Russia, Near East and Northern Africa and Western Europe. List of natural gas pipelines.

LNG carriers transport liquefied natural gas (LNG) across oceans, while tank trucks can carry liquefied or compressed natural gas (CNG) over shorter distances. Sea transport using CNG carrier ships that are now under development may be competitive with LNG transport in specific conditions.

Gas is turned into liquid at a liquefaction plant, and is returned to gas form at regasification plant at the terminal. Shipborne regasification equipment is also used. LNG is the preferred form for long distance, high volume transportation of natural gas, whereas pipeline is preferred for transport for distances up to 4,000 km over land and approximately half that distance offshore.

CNG is transported at high pressure, typically above 200 bars. Compressors and decompression equipment are less capital intensive and may be economical in smaller unit sizes than liquefaction/ regasification plants. Natural gas trucks and carriers may transport natural gas directly to end-users, or to distribution points such as pipelines.

In the past, the natural gas which was recovered in the course of recovering petroleum could not be profitably sold, and was simply burned at the oil field in a process known as flaring. Flaring is now illegal in many countries. Additionally, companies now recognise that gas may be sold to consumers in the form of LNG or CNG, or through other transportation methods. The gas is now re-injected into the formation for later recovery. The re-injection also assists oil pumping by keeping underground pressures higher.

A "master gas system" was invented in Saudi Arabia in the late 1970s, ending any necessity for flaring. Satellite observation, however, shows that flaring and venting are still practiced in some gas-extracting countries.

Natural gas is used to generate electricity and heat for desalination. Similarly, some landfills that also discharge methane gases have been set up to capture the methane and generate electricity.

Natural gas is often stored underground inside depleted gas reservoirs from previous gas wells, salt domes, or in tanks as liquefied natural gas. The gas is injected in a time of low demand and extracted when demand picks up. Storage nearby end users helps to meet volatile demands, but such storage may not always be practicable.

With 15 countries accounting for 84% of the worldwide extraction, access to natural gas has become an important issue in international politics, and countries vie for control of pipelines. In the first decade of the 21st century, Gazprom, the state-owned energy company in Russia, engaged in disputes with Ukraine and Belarus over the price of natural gas, which have created concerns that gas deliveries to parts of Europe could be cut off for political reasons.

Floating Liquefied Natural Gas (FLNG) is an innovative technology designed to enable the development of offshore gas resources that would otherwise remain untapped because due to environmental or economic factors it is nonviable to develop them via a land-based LNG operation. FLNG technology also provides a number of environmental and economic advantages:

Environmental – Because all processing is done at the gas field, there is no requirement for long pipelines to shore, compression units to pump the gas to shore, dredging and jetty construction, and onshore construction of an LNG processing plant, which significantly reduces the environmental footprint. Avoiding construction also helps preserve marine and coastal environments. In addition, environmental disturbance will be minimised during decommissioning because the facility can easily be disconnected and removed before being refurbished and re-deployed elsewhere.

Economic – Where pumping gas to shore can be prohibitively expensive, FLNG makes development economically viable. As a result, it will open up new business opportunities for countries to develop offshore gas fields that would otherwise remain stranded, such as those offshore East Africa.

Many gas and oil companies are considering the economic and environmental benefits of Floating Liquefied Natural Gas (FLNG). However, for the time being, the only FLNG facility now in development is being built by Shell, due for completion in around 2017.

Environmental Effects

CO_2 Emissions

Natural gas is often described as the cleanest fossil fuel, producing less carbon dioxide per joule delivered than either coal or oil and far fewer pollutants than other hydrocarbon fuels. However, in absolute terms, it does contribute substantially to global carbon emissions, and this contribution is projected to grow. According to the IPCC Fourth Assessment Report (Working Group III Report, chapter 4), in 2004, natural gas produced about 5.3 billion tons a year of CO_2 emissions, while coal and oil produced 10.6 and 10.2 billion tons respectively. According to an updated version of the SRESB emissions scenario, however, by the year 2030, natural gas would be the source of 11 billion tons a year, with coal and oil now 8.4 and 17.2 billion respectively because demand is increasing 1.9% a year. (Total global emissions for 2004 were estimated at over 27,200 million tons.)

In addition, natural gas itself is a greenhouse gas more potent than carbon dioxide. Although natural gas is released into the atmosphere in much smaller quantities, methane is oxidised in the atmosphere, and hence natural gas affects the atmosphere for approximately 12 years, compared to CO_2, which is already oxidised, and has effect for 100 to 500 years. Natural gas is composed mainly of methane, which has a radiative forcing twenty times greater than carbon dioxide. Based on such composition, a ton of methane in the atmosphere traps as much radiation as 20 tons of carbon dioxide; however, it remains in the atmosphere for 8–40 times less time. Carbon dioxide still receives the lion's share of attention concerning greenhouse gases because it is released in much larger amounts. Still, it is inevitable when natural gas is used on a large scale that some of it will leak into the atmosphere. (Coal methane not captured by coal bed methane extraction techniques is simply lost into the atmosphere. Current estimates by the EPA place global emissions of methane at 3 trillion cubic feet (85 km^3) annually, or 3.2% of global production. Direct emissions of methane represented 14.3% of all global anthropogenic greenhouse gas emissions in 2004.

Other Pollutants

Natural gas produces far lower amounts of sulfur dioxide and nitrous oxides than any other hydrocarbon fuel (fossil fuels). Carbon dioxide produced is 117,000 ppm vs 208,000 for burning coal. Carbon monoxide produced is 40 ppm vs 208 for burning coal. Nitrogen oxides produced is 92 ppm vs 457 for burning coal. Sulfur dioxide is 1 ppm vs 2,591 for burning coal. Mercury is 0 vs .016 for burning coal. Particulates are also a major contribution to global warming. Natural gas has 7ppm vs coal's 2,744ppm. Natural gas also has Radon, from 5 to 200,000 Becquerels per cubic metre.

Extraction

The practice of hydraulic fracturing, the process of using a combination of chemicals ranging from harmless to toxic to force natural gas to the surface from reservoirs with low permeability, has come under scrutiny internationally due to concerns about environmental and health safety, and has been suspended or banned in some countries. *Environmental concerns with hydraulic fracturing*

Safety Concerns

Figure: *A pipeline odourant injection station*

Production

In mines, where methane seeping from rock formations has no odour, sensors are used, and mining apparatus such as the Davy lamp has been specifically developed to avoid ignition sources.

Some gas fields yield sour gas containing hydrogen sulfide (H_2S). This untreated gas is toxic. Amine gas treating, an industrial scale process which removes acidic gaseouscomponents, is often used to remove hydrogen sulfide from natural gas.

Extraction of natural gas (or oil) leads to decrease in pressure in the reservoir. Such decrease in pressure in turn may result in subsidence, sinking of the ground above. Subsidence may affect ecosystems, waterways, sewer and water supply systems, foundations, and so on.

Releasing the gas from low-permeability reservoirs is accomplished by a process called hydraulic fracturing or "hydrofracking". To allow the natural gas to flow out of the shale, oil operators force 1 to 9 million US gallons (34,000 m^3) of water mixed with a variety of chemicals through the wellbore casing into the shale. The high pressure water breaks up or "fracks" the shale, which releases the trapped gas. Sand is added to the water as a proppant to keep the fractures in the shale open, thus enabling the gas to flow into the casing and then to the surface. The chemicals are added to the frack fluid to reduce friction and combat corrosion. During the extracting life of a gas well, other low concentrations of other chemical substances may be used, such as biocides to eliminate fouling, scale and corrosion inhibitors, oxygen scavengers to remove a source of corrosion, and acids to clean the perforations in the pipe.

Dealing with fracking fluid can be a challenge. Along with the gas, 30% to 70% of the chemically-laced frack fluid, or flow back, returns to the surface. Additionally, a significant amount of salt and other minerals, once a part of the rock layers that were under prehistoric seas, may be incorporated in the flow back as they dissolve in the frack fluid.

Use

In order to assist in detecting leaks, a minute amount of odourant is added to the otherwise colourless and almost odourless gas used by consumers. The odour has been compared to the smell of rotten eggs, due to the added butyl mercaptan. Sometimes a related compound, thiophane may be used in the mixture. Situations in which an odourant that is added to natural gas can be detected by analytical instrumentation, but cannot be properly detected by an observer with a normal sense of smell, have occurred in the natural gas industry. This is caused by odour masking, when one odourant overpowers the sensation of another. As of 2011, the industry is conducting research on the causes of odour masking.

Explosions caused by natural gas leaks occur a few times each year. Individual homes, small businesses and other structures are most frequently affected when an internal leak builds up gas inside the structure.

Frequently, the blast will be enough to significantly damage a building but leave it standing. In these cases, the people inside tend to have minor to moderate injuries.

Occasionally, the gas can collect in high enough quantities to cause a deadly explosion, disintegrating one or more buildings in the process. The gas usually dissipates readily outdoors, but can sometimes collect in dangerous quantities if flow rates are high enough. However, considering the tens of millions of structures that use the fuel, the individual risk of using natural gas is very low.

Natural gas heating systems are a minor source of carbon monoxide deaths in the United States. According to the US Consumer Product Safety Commission (2008), 56% of unintentional deaths from non-fire CO poisoning were associated with engine-driven tools like gas-powered generators and lawn mowers.

Natural gas heating systems accounted for 4% of these deaths. Improvements in natural gas furnace designs have greatly reduced CO poisoning concerns. Detectors are also available that warn of carbon monoxide and/or explosive gas (methane, propane, etc.).

Energy Content, Statistics, and Pricing

Quantities of natural gas are measured in normal cubic metres (corresponding to 0 °C at 101.325 kPa) or in standard cubic feet (corresponding to 60 °F (16 °C) and 14.73 psia). The gross heat of combustion of one cubic metre of commercial quality natural gas is around 39 megajoules (H 10.8 kWh), but this can vary by several percent. This comes to about 49 megajoules (H 13.5 kWh) for one kg of natural gas (assuming 0.8 kg/m^3, an approximate value).

The price of natural gas varies greatly depending on location and type of consumer. In 2007, a price of $7 per 1,000 cubic feet (28 m^3) was typical in the United States. The typical caloric value of natural gas is roughly 1,000 British thermal units (BTU) per cubic foot, depending on gas composition.

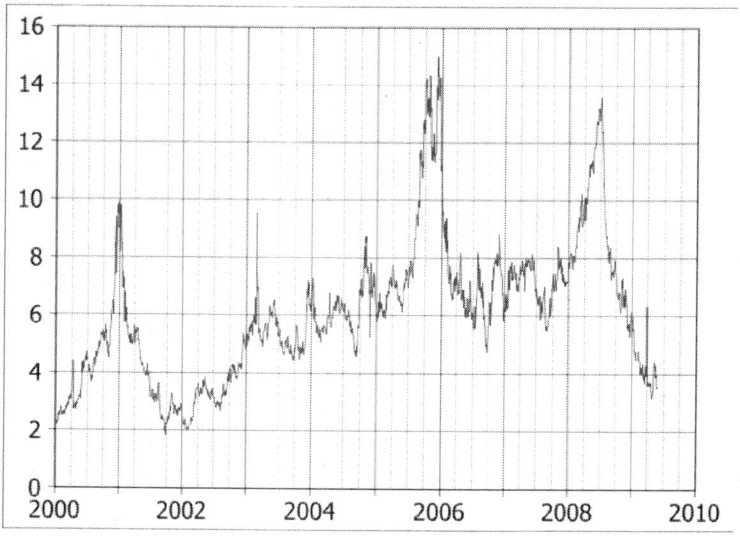

Figure: *Natural gas prices at the Henry Hub in US dollars per million BTUs ($/mmbtu) for 2000–2010.*

This corresponds to around $7 per million BTU, or around $7 per gigajoule. In April 2008, the wholesale price was $10 per 1,000 cubic feet (28 m^3) ($10/MMBTU). The residential price varies from 50% to 300% more than the wholesale price. At the end of 2007, this was $12–$16 per 1,000 cu ft (28 m^3). Natural gas in the United States is traded as a futures contract on the New York Mercantile Exchange. Each contract is for 10,000 MMBTU (~10,550 gigajoules), or 10 billion BTU. Thus, if the price of gas is $10 per million BTUs on the NYMEX, the contract is worth $100,000.

European Union

As one of the world's largest importers of natural gas, the EU is a major player on the international gas market.

Gas prices for end users vary greatly across the EU. A single European energy market, one of the key objectives of the European Union, should level the prices of gas in all EU member states.

United States

In US units, one standard cubic foot of natural gas produces around 1,028 British thermal units (BTU). The actual heating value when the water formed does not condense is the net heat of combustion and can be as much as 10% less.

In the United States, retail sales are often in units of therms (th); 1 therm = 100,000 BTU. Gas metres measure the volume of gas used, and this is converted to therms by multiplying the volume by the energy content of the gas used during that period, which varies slightly over time. Wholesale transactions are generally done in decatherms (Dth), or in thousand decatherms (MDth), or in million decatherms (MMDth). A million decatherms is roughly a billion cubic feet of natural gas. Gas sales to domestic consumers may be in units of 100 standard cubic feet (Ccf).

As of 2009, the Potential Gas Committee estimated that the United States has total future recoverable natural gas resources approximately 100 times greater than current annual consumption.

Canada

Canada uses metric measure for internal trade of petrochemical products. Consequently, natural gas is sold by the Gigajoule, a measure approximately equal to 1/2 of a barrel (250lbs) of oil, or 1 million BTUs, or 1000 cu ft of gas, or 28cu metres of gas.

Elsewhere

In the rest of the world, natural gas is sold in Gigajoule retail units. LNG (liquefied natural gas) and LPG (liquefied petroleum gas) are traded in metric tons or mmBTU as spot deliveries. Long term natural gas distribution contracts are signed in cubic metres, and LNG contracts are in metric tonnes (1,000kg). The LNG and LPG is transported by specialised transport ships, as the gas is liquified at cryogenic temperatures. The specification of each LNG/LPG cargo will usually contain the energy content, but this information is in general not available to the public. In the Russian Federation, Gazprom sold approximately 250 billion cubic metres of natural gas in 2008.

Petroleum Licensing

Petroleum licensing or exploration license is the act of attributing licenses (geographical areas at land and/or sea) to a company or a Join Venture, for the act of petroleum exploration.

Each sovereign country distributes licenses in what is typically called a licensing round. The procedure for such, can greatly vary from country to country. The largest change is usually if it is a bid system or a grant system. In the bid system, each company or joint venture will offer a bid to gain the rights for the petroleum exploration at the license for a limited period of time. The highest bid, will obtain the rights. In the grant system, the license will be granted to the company or join venture that shows the highest interest and ability for the exploration of the license. This can be shown by the company experience, projected plan for the exploration (by high investment) and or longest presence on the country as a petroleum exploration company.

Countries Licensing Rounds

- United Kingdom - The Department of Energy and Climate Change (DECC - formerly the Department of Trade and Industry) grants licences. The UKCS (United Kingdom Continental Shelf) is divided into quadrants of 1 degree latitude and one degree longitude. Each quadrant is divided into 30 blocks measuring 10 minutes of latitude and 12 minutes of longitude. Some blocks are divided further into part blocks where some areas are relinquished by previous licensees. For example, block 13/24a is located in quad 13 and is the 24th block and is the 'a' part block. The UK government has traditionally issued licences via periodic (now annual) licensing rounds. Blocks are awarded on the basis of the work programme bid by the participants. The UK government has actively solicited new entrants to the UKCS via "promote" licensing rounds with less demanding terms and the fallow acreage initiative, where non-active licences have to be relinquished.
- Norway - The Norwegian Petroleum Directorate grants licences. The NCS is also divided into quads of 1 degree by 1 degree. Norwegian licence blocks are larger than British blocks, being 15 minutes of latitude by 20 minutes of longitude (12 blocks in a quad). Like in Britain, there are numerous part blocks formed by re-licensing relinquished areas. One of the attractiveness of the license blocks in Norway are the economical benefits given by its taxation policy. The Norwegian tax system refunds 78% of all directly related exploration costs, which are paid the year after the costs occurred.
- Denmark - The Danish Energy Authority administers the Danish sector. The Danes also divide their sector of the North Sea into 1 degree by 1 degree quadrants. Their blocks, however, are 10 minutes latitude by 15 minutes longitude. Part blocks exist where partial relinquishments have taken place.
- Germany - Germany and the Netherlands share a quadrant and block grid - quadrants are given letters rather than numbers. The blocks are 10 minutes latitude by 20 minutes longitude. Germany has the smallest sector in the North Sea.
- Brazil - The National Petroleum Agency grants licenses. The license blocks are named by BT or BM, followed by the state

code and a number. BT means the block is onshore, on land. While BM means the block is offshore, at sea.

Relinquishment

When a petroleum license is granted, the company or join venture is given a limited time for the exploration of the license. If after the limited time, the company has not discovered or performed its minimal obligations (usually agreed with the government as a minimum number of exploration wells and investment on seismic) the license will be relinquished.

Relinquished licenses can be re-attributed in next licensing rounds, or even kept as relinquished if the government believes there is little interest on that area for the petroleum exploration.

Chapter 7

Waste Oil

Waste oil is defined as any petroleum-based or synthetic oil that, through use or handling, has become unsuitable for its original purpose due to the presence of impurities or loss of original properties.

Differentiating between "Waste Oil" and "Used Oil"

The U.S. EPA defines the term "used oil" using (approximately) the definition given above, while the term "waste oil" is used to mean used oil (or oil) that is inappropriately mixed with hazardous waste, or disposed of, rather than recycled (re-refined, burned for energy recovery, etc.). The general difference between the two terms is essentially whether the used oil is disposed of, or appropriately recycled.

Products Used as Waste Oil

Some examples of types of products that after use, can be labelled as used oil are: hydraulic oil, transmission oil, brake fluids, motor oil, crankcase oil, gear box oil, synthetic oil, and grades 1, 2, 3 and 4 fuel oil.

Disposal of Waste Oil

Waste oil can be disposed of in different ways, including sending the used oil off-site (some facilities are permitted to handle the used oil such as your local garages and local waste disposal facilities), burning used oil as a fuel (some used oil is not regulated by burner standards, but others that are off-specification used oil can only be burned in either industrial furnaces, certain boilers, and permitted hazardous waste incinerators), and marketing the used oil (claims are made that the used oil is to be burned for energy recovery, it is then shipped to a used oil burner who burns the used oil in an approved industrial furnace or boiler). Oils that are off-specification typically contain: Arsenic 5 ppm, Cadmium 2 ppm, Chromium 10 ppm, Lead 100 ppm, Flash point 100°F, minimum (i.e., fp must be greater than 100°F), Total Halogens >4,000 ppm

Storage and Handling of Waste Oil

For on-site burning of used oil, the oil must be stored in tanks or containers, above or underground. The containers must be in good condition with no leaks, the tanks/containers must be labelled with the words "used oil", and there must be a spill prevention plan (or a control and countermeasures plan).

Waste Oil Furnaces and Boilers

Waste oil furnace is a type of furnace used for heating purposes and is fueled by used oil that is free of hazardous contaminants, as described by the EPA. Waste-oil-fueled boilers can be used for various industrial purposes as well as heating.

Dangerous Goods

Dangerous goods are solids, liquids, or gases that can harm people, other living organisms, property, or the environment. They are often subject to chemical regulations. "HazMat teams" are personnel specially trained to handle dangerous goods. Dangerous goods include materials that are radioactive, flammable, explosive, corrosive, oxidizing, asphyxiating, biohazardous, toxic, pathogenic, or allergenic. Also included are physical conditions such as compressed gases and liquids or hot materials, including all goods containing such materials or chemicals, or may have other characteristics that render them hazardous in specific circumstances.

Dangerous goods are often indicated by diamond-shaped signage. The colours of each diamond in a way has reference to its hazard i.e.: Flammable = red because fire and heat are generally of red colour, Explosive = orange, because mixing red (flammable) with yellow (oxidising agent) creates orange. Non Flammable Non Toxic Gas = green, due to all compressed air vessels being this colour in France after World War II. France is where the diamond system of HazMat identification originated.

Mitigating the risks associated with hazardous materials may require the application of safety precautions during their transport, use, storage and disposal. Most countries regulate hazardous materials by law, and they are subject to several international treaties as well. Even so, different countries may use different class diamonds for the same product. For example, in Australia, Anhydrous Ammonia UN

1005 is classified as 2.3 (Toxic Gas) with sub risk 8 (Corrosive), whereas in the U.S. it is only classified as 2.2 (Non Flammable Gas).

People who handle dangerous goods will often wear protective equipment, and metropolitan fire departments often have a response team specifically trained to deal with accidents and spills. Persons who may come into contact with dangerous goods as part of their work are also often subject to monitoring or health surveillance to ensure that their exposure does not exceed occupational exposure limits.

Laws and regulations on the use and handling of hazardous materials may differ depending on the activity and status of the material. For example, one set of requirements may apply to their use in the workplace while a different set of requirements may apply to spill response, sale for consumer use, or transportation. Most countries regulate some aspect of hazardous materials.

The most widely applied regulatory scheme is that for the transportation of dangerous goods. The United Nations Economic and Social Council issues the *UN Recommendations on the Transport of Dangerous Goods*, which form the basis for most regional and national regulatory schemes. For instance, the International Civil Aviation Organisation has developed regulations for air transport of hazardous materials that are based upon the UN Model but modified to accommodate unique aspects of air transport. Individual airline and governmental requirements are incorporated with this by the International Air Transport Association to produce the widely used IATA *Dangerous Goods Regulations* (DGR).

Similarly, the International Maritime Organisation has developed the International Maritime Dangerous Goods Code ("IMDG Code", part of the International Convention for the Safety of Life at Sea) for transportation on the high seas, and the Intergovernmental Organisation for International Carriage by Rail has developed the Regulations concerning the International Carriage of Dangerous Goods by Rail ("RID", part of the Convention concerning International Carriage by Rail). Many individual nations have also structured their dangerous goods transportation regulations to harmonize with the UN Model in organisation as well as in specific requirements.

The Globally Harmonized System of Classification and Labelling of Chemicals (GHS) is an internationally agreed upon system set to replace the various classification and labelling standards used in different countries. GHS will use consistent criteria for classification and labelling on a global level. Dangerous goods are divided into classes on the basis of the specific chemical characteristics producing the risk.

Not all countries use precisely the same graphics (label, placard and/or text information) in their national regulations. Some use graphic symbols, but without English wording or with similar wording in their national language. Refer to the Dangerous Goods Transportation Regulations of the country of interest.

The statement above applies equally to all the Dangerous Goods classes discussed in this article.

Australia

The Australian Dangerous Goods Code, seventh ion (2008) complies with international standards of importation and exportation of dangerous goods in line with the UN Recommendations on the Transport of Dangerous Goods. Australia uses the standard international UN numbers with a few slightly different signs on the back, front and sides of vehicles carrying hazardous substances. The country uses the same "Hazchem" code system as the UK to provide advisory information to emergency services personnel in the event of an emergency.

Canada

Transportation of dangerous goods (hazardous materials) in Canada by road is normally a provincial jurisdiction. The federal government has jurisdiction over air, most marine, and most rail transport. The federal government acting centrally created the federal transportation of dangerous goods act and regulations, which provinces adopted in whole or in part via provincial transportation of dangerous goods legislation. The result is that all provinces use the federal regulations as their standard within their province; some small variances can exist because of provincial legislation. Creation of the federal regulations was coordinated by Transport Canada. Hazard classifications are based upon the UN Model.

The province of Nova Scotia's dangerous goods transportation act can be viewed here: Dangerous Goods Transportation Act

The province of Nova Scotia's dangerous goods transportation regulations can can be viewed here: Dangerous Goods Transportation Regulations. Outside of federal facilities, labour standards are generally under the jurisdiction of individual provinces and territories. However, communication about hazardous materials in the workplace has been standardised across the country through Health Canada's Workplace Hazardous Materials Information System (WHMIS).

Europe

The European Union has passed numerous directives and regulations to avoid the dissemination and restrict the usage of hazardous substances, important ones being the Restriction of Hazardous Substances Directive and the REACH regulation. There are also long-standing European treaties such as ADR and RID that regulate the transportation of hazardous materials by road, rail, river and inland waterways, following the guide of the UN Model Regulation.

European law distinguishes clearly between the law of dangerous goods and the law of hazardous materials. The first refers primarily to the transport of the respective goods including the interim storage, if caused by the transport. The latter describes the requirements of storage (including warehousing) and usage of hazardous materials. This distinction is important, because different directives and orders of European law are applied.

United Kingdom

The United Kingdom (and also Australia, Malaysia, and New Zealand) use the Hazchem warning plate system which carries information on how an emergency service should deal with an incident. The *Dangerous Goods Emergency Action Code List* (EAC) lists dangerous goods; it is reviewed every two years and is an essential compliance document for all emergency services, local government and for those who may control the planning for, and prevention of, emergencies involving dangerous goods.

United States

Due to the increase in the perceived threat of terrorism in the early 21st century after the September 11, 2001 attacks, funding for greater HAZMAT-handling capabilities was increased throughout the United States, recognising that flammable, poisonous, explosive, or radioactive substances in particular could be used for terrorist attacks.

The United States Department of Transportation (DOT) regulates hazmat transportation within the territory of the US by *Title 49 of the Code of Federal Regulations*. The U.S. Occupational Safety and Health Administration (OSHA) regulates the handling of hazardous materials in the workplace as well as response to hazardous-materials-related incidents, most notably through *Hazardous Waste Operations and*

Emergency Response (*HAZWOPER*). regulations found at 29 FR 1910.120.

The Environmental Protection Agency (EPA) regulates hazardous materials as they may impact the community and environment, including specific regulations for environmental cleanup and for handling and disposal of waste hazardous materials.

The Consumer Product Safety Commission regulates hazardous materials that may be used in products sold for household and other consumer uses.

Hazard Classes for Materials in Transport

Following the UN Model, the DOT divides regulated hazardous materials into nine classes, some of which are further subdivided. Hazardous materials in transportation must be placarded and have specified packaging and labelling. Some materials must always be placarded, others may only require placarding in certain circumstances.

Trailers of goods in transport are usually marked with a four digit UN number. This number can be referenced by first responders (firefighters, police officers, and ambulance personnel) who can find information about the material in the *Emergency Response Guidebook*.

Fixed Facilities

Different standards usually apply for handling and marking HAZMATs at fixed facilities, including NFPA 704 diamond markings (a consensus standard often adopted by local governmental jurisdictions), OSHA regulations requiring chemical safety information for employees, and CPSC requirements requiring informative labelling for the public, as well as wearing Hazmat suits when handling hazardous materials.

Packing Groups

Packing groups are used for the purpose of determining the degree of protective packaging required for Dangerous Goods during transportation.

- Group I: great danger, and most protective packaging required. Some combinations of different classes of dangerous goods on the same vehicle or in the same container are forbidden if one of the goods is Group I.
- Group II: medium danger
- Group III: least danger among regulated goods, and least protective packaging within the transportation requirement

Transport Documents

One of the transport regulations is that, as an assistance during emergency situations, written instructions how to deal in such need to be carried and easily accessible in the driver's cabin. A license or permit card for hazmat training must be presented when requested by officials. Dangerous goods shipments also require a special declaration form prepared by the shipper.

Among the information that is generally required includes the shipper's name and address; the consignee's name and address; descriptions of each of the dangerous goods, along with their quantity, classification, and packaging; and emergency contact information. Common formats include the one issued by the International Air Transport Association (IATA) for air shipments and the form by the International Maritime Organisation (IMO) for sea cargo.

Oil Depletion

Oil depletion occurs in the second half of the production curve of an oil well, oil field, or the average of total world oil production. The Hubbert peak theory makes predictions of production rates based on prior discovery rates and anticipated production rates. Hubbert curves predict that the production curves of non-renewing resources approximate a bell curve. Thus, when the peak of production is passed, production rates enter an exponential decline.

The American Petroleum Institute estimated in 1999 the world's oil supply would be depleted between 2062 and 2094, assuming total world oil reserves at between 1.4 and 2 trillion barrels (220 and 320 km^3) and consumption at 80 million barrels per day (13,000,000 m^3/d). In 2004, total world reserves were estimated to be 1.25 trillion barrels (199 km^3) and daily consumption was about 85 million barrels (13,500,000 m^3), shifting the estimated oil depletion year to 2057. A study published in the journal Energy Policy by researchers from Oxford University, however, predicted demand would surpass supply by 2015 (unless constrained by strong recession pressures caused by reduced supply or government intervention).

The United States Energy Information Administration predicted in 2006 that world consumption of oil will increase to 98.3 million barrels per day (15,630,000 m^3/d) (mbd) in 2015 and 118 mbd in 2030. With 2009 world oil consumption at 84.4 mbd, reaching the projected 2015 level of consumption would represent an average annual increase between 2009 and 2015 of 2.7% per year while EIA's declining consumption and declining supplies during the 2005-2009 period.

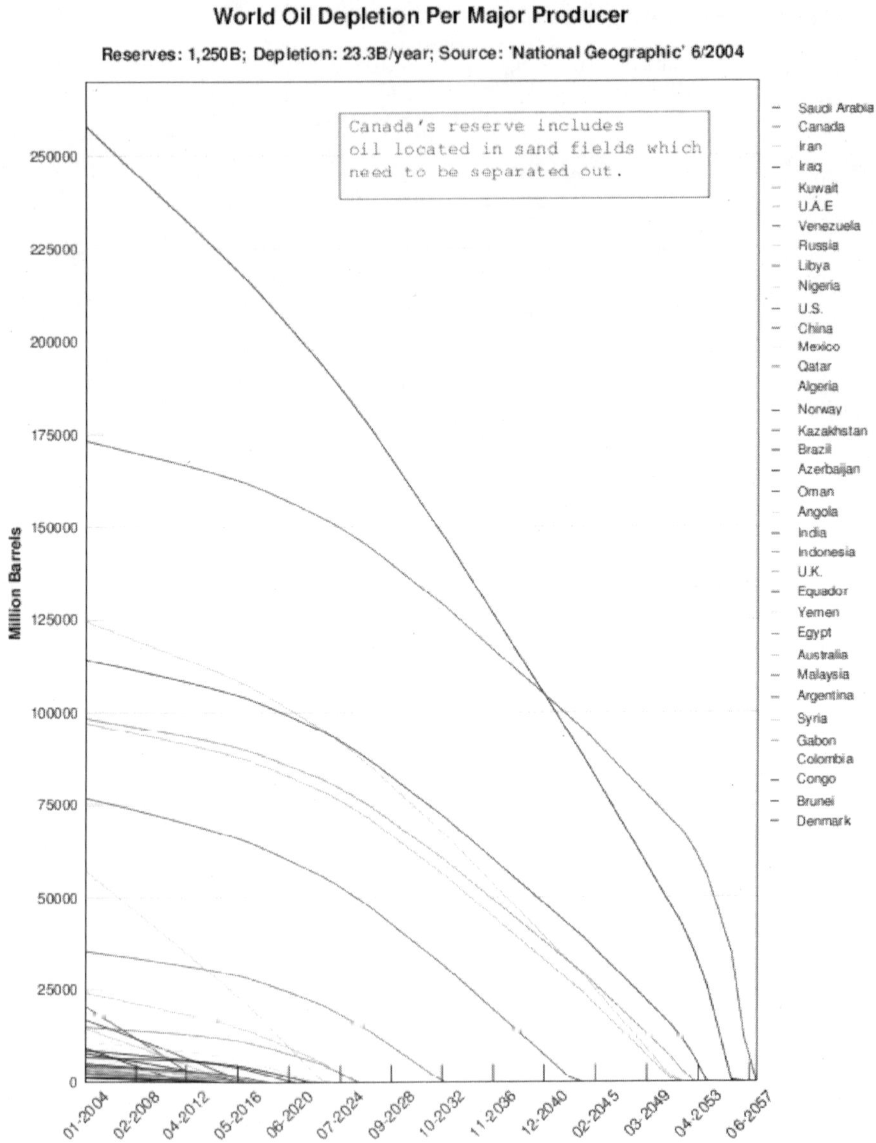

Figure : *Oil depletion per major producing country. This model assumes world oil depletion remains constant at the 2004 level of 80 million barrels per day. However, world oil depletion is currently (as of 2008) at 85 million barrels per day. This model also assumes production will increase in the high-capacity countries as low-capacity countries become depleted.*

Resource Availability

The world's oil supply is fixed because petroleum is naturally formed far too slowly to be replaced at the rate at which it is being extracted. Over many millions of years, plankton, bacteria, and other plant and animal matter become buried in sediments on the ocean floor. When conditions are right – a lack of oxygen for decomposition, and sufficient depth and temperature of burial – these organic remains are converted into petroleum compounds, while the sediment accompanying them is converted into sandstone, siltstone, and other porous sedimentary rock. When capped by impermeable rocks such as shale, salt, or igneous intrusions, they form the petroleum reservoirs which are exploited today.

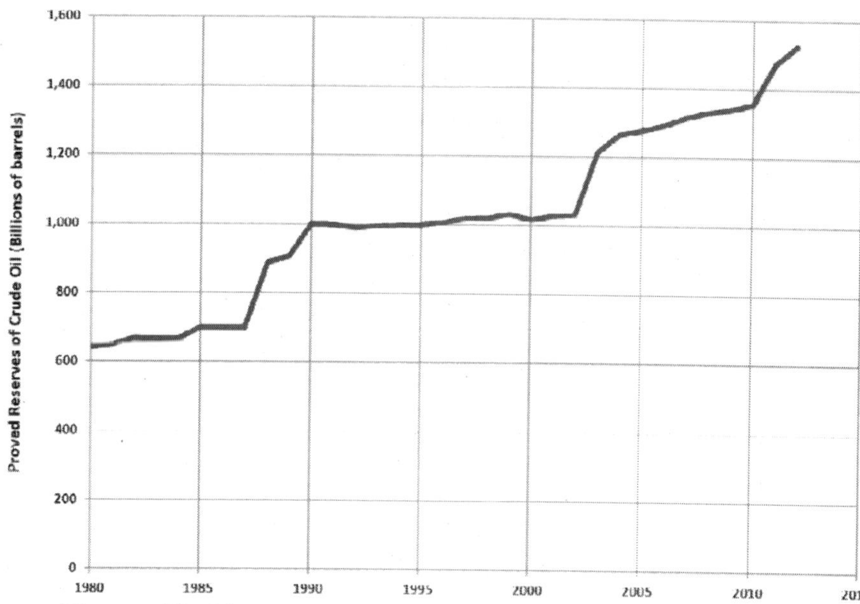

Figure: *World proved reserves of crude oil, 1980-2012 (US EIA)*

Production Decline Models

Oil production decline occurs in a predictable manner based on geological circumstances, governmental policies, and engineering practices. The shape of the decline curve varies depending upon whether one considers a well, a field, a set of fields, or the world.

Oil Well Production Decline

Oil well production curves typically end in an exponential decline. At natural rates, oil well production curves appear similar to a bell curve, a phenomenon known as the Hubbert curve. The typical decline is a rapid drop in production, and eventually a levelling off to a point at which they no longer produce profitable amounts. Such wells are referred to as marginal or stripper wells.

The shape of production curve of an oil well can be affected by a number of factors:

- Well may be restricted by choice by lack of market demand or government regulation. This flattens the peak of the curve, but will not change the well's total production significantly.
- Hydraulic fracturing (*fracking*) or acidizing may be used to cause a sharp spike in production, and may increase the recoverable reserves of a given well.
- The field may undergo a secondary or tertiary recovery project, discussed in the next section.

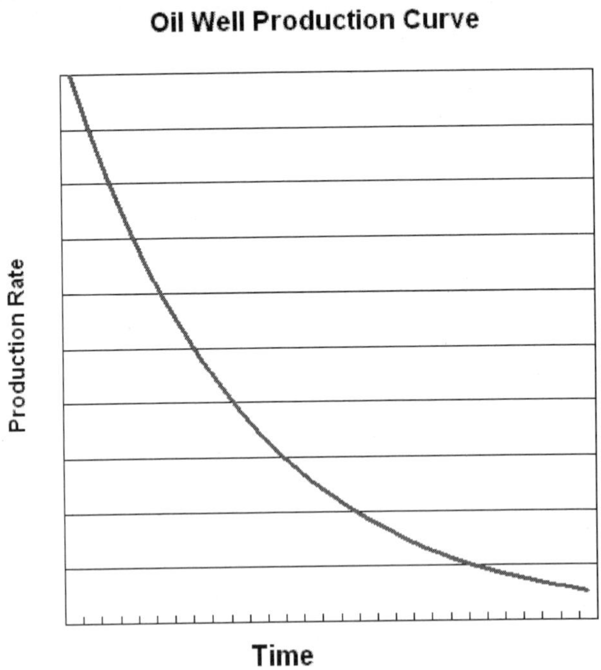

Figure: *Theoretical oil production curve for a well with exponential decline*

Oil Field Production Decline

Each individual oil well is a portion of a larger fixed area oil field. As with individual wells, discovery and production amounts of oil fields generally average to a similar bell shaped production curve. Eventually, when the field is completely drilled out, a field's production goes into a sharp decline as the average production of its wells enter decline. As this decline levels off, production can continue at relatively low rates. A number of oil fields in the U.S. have been producing for over 100 years. Oil field production curves can be modified by a number of factors:

- Production may be restricted by market conditions or government regulation.
- A secondary recovery project, such as water or gas injection, can repressurise the field and improve the production rate temporarily. However, it will not change the total production amount over the life of the field. Eventually the field will go into a steeper than normal decline.
- the field may undergo an enhanced oil recovery project, such as drilling of wells for injection of solvents, carbon dioxide, or steam. This can be very expensive but allows more oil to be coaxed out of the rock, increasing the ultimate production of the field.

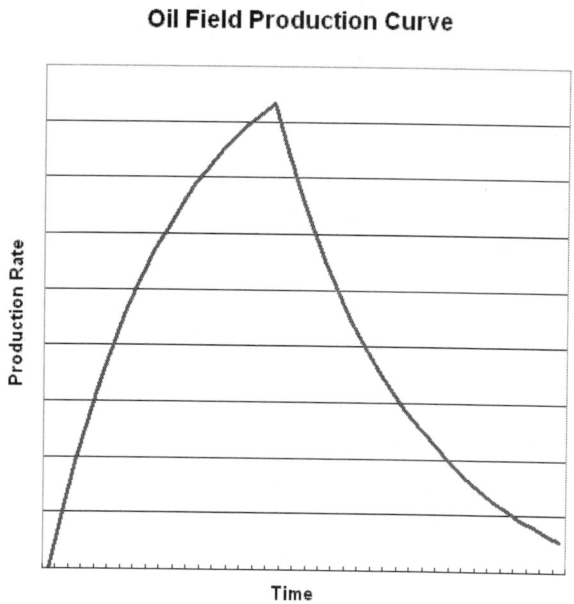

Figure: *Typical oil field production curve*

Multi-field Production Decline

Most oil is found in a small number of very large oil fields. If oil fields are discovered at a constant rate until they have all been found, the combined production of fields will yield a curve such as the one at right. Production starts off slowly, rises faster and faster, then slows down and flattens until it reaches a peak. After the production peak, production enters an exponential decline, eventually flattening out. Oil production may never actually reach zero, but eventually becomes very low. Factors which can modify this curve include:

- Inadequate demand for oil, which reduces steepness of the curve and pushes its peak into the future.
- Sharp price increases when the production peak is reached, as production fails to meet demand. If price increases cause a sharp drop in demand, a dip in the top of the curve may occur.
- Development of new drilling technology or marketing of unconventional oil can reduce the steepness of the decline as more oil is produced than initially anticipated.

Multiple Oil Field Production Curve

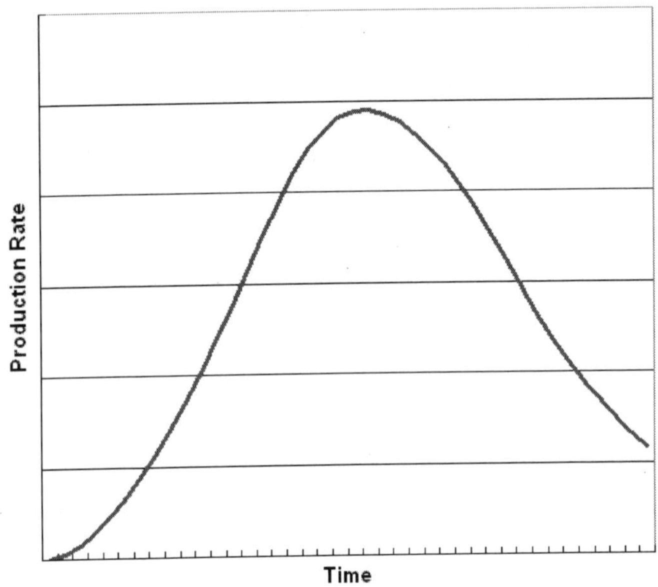

Figure: *Multiple oil field production curve*

United States Production Decline

Oil production in the United States has followed the theoretical Hubbert curve. U.S. oil production reached its peak in 1970 and by the mid-2000s it had fallen to 1940s levels. In 1950, the United States produced over half the world's oil, but by 2005 that proportion had dropped to about 8%. In 2005, U.S. crude oil imports were twice as high as domestic production. The production peak in 1970 was predicted in 1956 by Hubbert. By 1972 all import quotas and controls on U.S. domestic production had been removed. Oil companies began drilling large numbers of oil wells on a nationwide scale. Despite this, and despite the quadrupling of prices during the 1973 oil crisis, the production decline has, to date, proven irreversible.

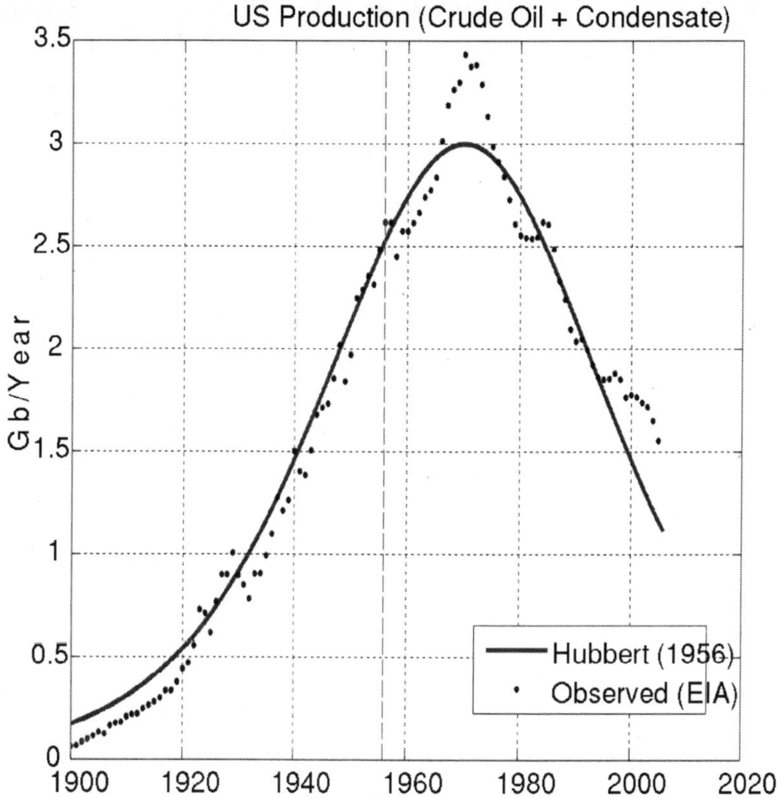

Figure: *U.S. oil production in the lower 48 states (crude oil only) to 2005 and Hubbert's high estimate (a multi-field projection)*

The actual U.S. production curve does deviate from Hubbert's 1956 curve in some significant ways:

- When oil surpluses created a glut on the market and low prices began causing demand and production curves to rise, regulatory agencies such as the Texas Railroad Commission stepped in to restrain production.
- The curve peaked at a sharp point rather than gradually flattening out. This occurred because as consumption began to approach production limits, oil companies drilled out all their existing fields as fast as they could, and many of those fields peaked simultaneously.
- Production fell after 1970, but started to recover and reached a lower secondary peak in 1988. This occurred because the supergiant Prudhoe Bay field in Alaska was only discovered in 1968, and the Trans-Alaska Pipeline System (TAPS) was not

completed until 1977. After 1988, Alaska production peaked and total U.S. production began to decline again. By 2005, Prudhoe Bay had produced over 75% of its oil.

World Oil Production

World oil production has followed a typical Hubbert curve, rising over the past century with only a few dips. The 1970 production peak in the U.S. caused many people to begin to question when the world production peak would occur. The peak of world production is known as Peak oil. By the mid-2000s, all of the world's major oil producing countries except Saudi Arabia were producing at maximum capacity (many having peaked in production), and some experts such as Matthew Simmons were questioning whether even Saudi Arabia had any reserve capacity left.

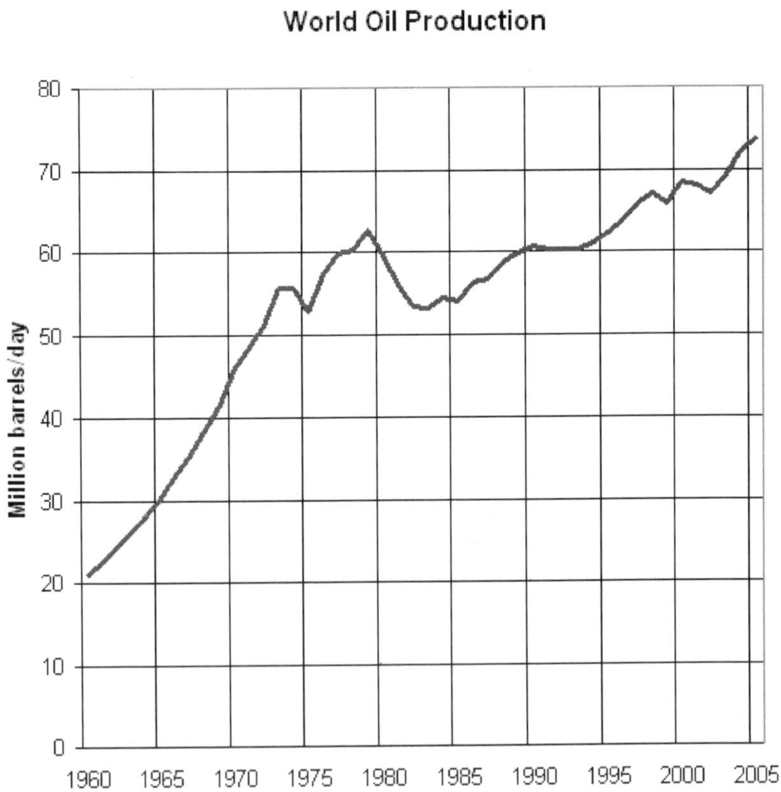

Figure: *World oil field production curve*

Industry observers have pointed to the similarities between the global production curve in mid-2000s and that of the United States in the 1970s.

- The oil price increases since 2003 were preceded by a decade of production cutbacks in OPEC countries in an attempt to keep prices high despite an oil glut. This is similar to production cutbacks in Texas and other states to maintain prices despite an oil glut in the decade prior to the 1973 oil crisis.
- World oil prices reached record inflation adjusted highs beginning in 2008, but new oil did not appear on the market, as the theory of supply and demand would predict. This is reminiscent of price increases in the United States in the 1970s when U.S. oil production started to decline despite record high prices and record drilling by oil companies.
- There are serious doubts about whether OPEC countries really have the oil reserves they claim. This is similar to the illusionary oil reserves that U.S. oil companies claimed to have in the decade prior to the 1973 and 1979 oil crisis. In the 1970s, those companies were unable to produce as much oil as they had predicted, and production went down instead of up.

Implications of a World Peak

A peak in oil production could result in a worldwide oil shortage, or it could not even be noticed as demand decreases in conjunction with increased prices. While past shortages stemmed from a temporary insufficiency of supply, crossing Hubbert's Peak would mean that the production of oil would continue to decline, and that demand for these products must be reduced to meet supply. The effects of such a shortage would depend on the rate of decline and the development and adoption of effective alternatives.

If alternatives were not forthcoming, it has been speculated that the numerous products produced with oil would become scarcer, leading to at the very least lower living standards in developed and developing countries alike, and possibly in the worst case to the collapse of the entire international banking system, which could not likely sustain itself without the prospect of growth. The political situation may change dramatically, with potential wars between countries over access to dwindling supplies. Accordingly, inequalities between various countries and regions of the world may become exacerbated.

Catastrophe

Economic growth and prosperity since the industrial revolution have, in large part, been due to increased efficiencies in the use of better and higher concentrations of energy in fossil fuels. The use of fossil fuels allows humans to participate in takedown, which is the consumption of energy at a greater rate than it is being replaced. Some believe that decreasing oil production portends a drastic impact on human culture and modern technological society, which is currently heavily dependent on oil as a fuel and chemical feedstock. For example, over 90% of transportation in the United States relies on oil.

Some envisage a Malthusian catastrophe occurring as oil becomes increasingly inefficient to produce; others have learned from the examples demonstrated in mature basins and applied those operational procedures to these basins to preserve their operational tempo. Since the 1940s, agriculture has dramatically increased its productivity, due largely to the use of chemical pesticides, fertilizers, and increased mechanisation. This process has been called the Green Revolution. The increase in food production has allowed world population to grow dramatically over the last 50 years. Pesticides rely upon oil as a critical ingredient, and fertilizers require natural gas. Farm machinery also requires oil.

Arguing that in today's world every joule one eats requires 5–15 joules to produce and deliver, some have speculated that decreasing supply of oil will cause modern industrial agriculture to collapse, leading to a drastic decline in food production, food shortages and possibly even mass starvation. However, most or all of the uses of fossil fuels in agriculture can be replaced with alternatives. For example, by far the biggest fossil fuel input to agriculture is the use of natural gas as a hydrogen source for the Haber-Bosch fertilizer-creation process. Natural gas is used simply because it is the cheapest currently-available source of hydrogen; were that to change, other sources, such as electrolysis powered by solar energy, could be used to provide the hydrogen for creating fertilizer without relying on fossil fuels.

Oil shortages may force a move to lower input "organic agriculture" methods, which may be more labour-intensive and require a population shift from urban to rural areas, reversing the trend towards urbanisation which has predominated in industrial societies; however, some organic farmers using modern organic-farming methods have reported yields as high as those available from conventional farming, but without the use of fossil-fuel-intensive artificial fertilizers or pesticides.

Another possible effect would derive from modern transportation and housing infrastructure. A large proportion of the developed world's population live in suburbs, a type of low-density settlement designed with the automobile in mind. Commentators such as James Howard

Kunstler argue that because of its reliance on the automobile, the suburb is an unsustainable living arrangement; the implications of peak oil would leave many suburb-dwellers unable to afford fuel for their cars, and force them to move to higher density, more walkable areas. In effect, surburbia would comprise the "slums of the future." A movement to deal with this problem early, called "New Urbanism," seeks to develop the suburbs into higher density neighbourhoods and use high density, mixed-use forms for new building projects.

Recession

A more modest scenario, assuming a slower rate of depletion and a smooth transition to alternative energy sources could cause substantial economic hardship such as a recession or depression due to higher energy prices. Historically, there is a close correlation in the timing of oil price spikes and economic downturns. Inflation has also been linked to oil price spikes.

However, economists disagree on the strength and causes of this association. Conversely, the recessions of the early 1970s and early 1980s were associated with a relatively brief period of somewhat dwindling energy availability; the possible future increase in oil prices might be much higher and last longer.

Rising Food Prices

Rising oil prices cause rising food prices in three ways. First, increased equipment fuel costs drive higher prices. Second, transportation costs increase retail prices. Third, higher oil prices are causing farmers to switch from producing food crops to producing biofuel crops. The law of supply and demand predicts that if fewer farmers are producing food the price of food will rise.

Motor Oil

Motor oil or engine oil is an oil used for lubrication of various internal combustion engines. The main function is to lubricate moving parts; it also cleans, inhibit scorrosion, improves sealing, and cools the engine by carrying heat away from moving parts.

Motor oils are derived from petroleum-based and non-petroleum-synthesized chemical compounds. Motor oils today are mainly blended by using base oils composed of hydrocarbons, polyalphaolefins (PAO), and polyinternal olefins (PIO), thus organic compounds consisting entirely of carbon and hydrogen. The base oils of some high-performance motor oils contain up to 20% by weight of esters.

Use

Motor oil is a lubricant used in internal combustion engines. These include motor or road vehicles such as cars and motorcycles, heavier vehicles such as buses and commercial vehicles, non-road vehicles such as go-karts, snow mobiles, boats (fixed engine installations and outboards), lawn mowers, large agricultural and construction equipment, locomotives and aircraft and static engines such as electrical generators. In engines, there are parts which move against each other causing friction which wastes otherwise useful power by converting the energy to heat. Contact between moving surfaces also wears away those parts, which could lead to lower efficiency and degradation of the engine. This increases fuel consumption, decreases power output and can, in extreme cases lead to engine failure.

Lubricating oil creates a separating film between surfaces of adjacent moving parts to minimise direct contact between them, decreasing heat caused by friction and reducing wear, thus protecting the engine. In use, motor oil transfers heat through convection as it flows through the engine by means of air flow over the surface of the oil pan, an oil cooler and through the build up of oil gases evacuated by the Positive Crankcase Ventilation (PCV) system.

In petrol (gasoline) engines, the top piston ring can expose the motor oil to temperatures of 160 °C (320 °F). In diesel engines the top ring can expose the oil to temperatures over 315 °C (600 °F). Motor oils with higher viscosity indices thin less at these higher temperatures.

Coating metal parts with oil also keeps them from being exposed to oxygen, inhibiting oxidation at elevated operating temperatures preventing rust or corrosion. Corrosion inhibitors may also be added to the motor oil. Many motor oils also have detergents and dispersants added to help keep the engine clean and minimise oil sludge build-up. The oil is able to trap soot from combustion in itself, rather than leaving it deposited on the internal surfaces. It is a combination of this, and some singeing that turns used oil black after some running.

Rubbing of metal engine parts inevitably produces some microscopic metallic particles from the wearing of the surfaces. Such particles could circulate in the oil and grind against moving parts, causing wear. Because particles accumulate in the oil, it is typically circulated through an oil filter to remove harmful particles. An oil pump, a vane or gear pump powered by the engine, pumps the oil throughout the engine, including the oil filter. Oil filters can be a *full flow* or *bypass* type.

In the crankcase of a vehicle engine, motor oil lubricates rotating or sliding surfaces between the crankshaft journal bearings (main bearings and big-end bearings), and rods connecting the pistons to the crankshaft. The oil collects in an oil pan, or sump, at the bottom of the crankcase. In some small engines such as lawn mower engines, dippers on the bottoms of connecting rods dip into the oil at the bottom and splash it around the crankcase as needed to lubricate parts inside. In modern vehicle engines, the oil pump takes oil from the oil pan and sends it through the oil filter into oil galleries, from which the oil lubricates the main bearings holding the crankshaft up at the main journals and camshaft bearings operating the valves. In typical modern vehicles, oil pressure-fed from the oil galleries to the main bearings enters holes in the main journals of the crankshaft. From these holes in the main journals, the oil moves through passageways inside the crankshaft to exit holes in the rod journals to lubricate the rod bearings and connecting rods. Some simpler designs relied on these rapidly moving parts to splash and lubricate the contacting surfaces between the piston rings and interior surfaces of the cylinders. However, in modern designs, there are also passageways through the rods which carry oil from the rod bearings to the rod-piston connections and lubricate the contacting surfaces between the piston rings and interior surfaces of the cylinders. This oil film also serves as a seal between the piston rings and cylinder walls to separate the combustion chamber in the cylinder head from the crankcase. The oil then drips back down into the oil pan.

Non-vehicle Motor Oils

An example is lubricating oil for four-stroke or four-cycle internal combustion engines such as those used in portable electricity generators and "walk behind" lawn mowers. Another example is two-stroke oil for lubrication of two-stroke or two-cycle internal combustion engines found in snow blowers, chain saws, model aeroplanes, gasoline powered gardening equipment like hedge trimmers, leaf blowers and soil cultivators. Often, these motors are not exposed to as wide service temperature ranges as in vehicles, so these oils may be single viscosity oils.

In small two-stroke engines, the oil may be pre-mixed with the gasoline or fuel, often in a rich gasoline: oil ratio of 25:1, 40:1 or 50:1, and burned in use along with the gasoline. Larger two-stroke engines used in boats and motorcycles, will have a more economical oil injection system rather than oil pre-mixed into the gasoline. The oil injection system is not used on small engines used in applications like snowblowers and trolling motors as the oil injection system is too expensive for small engines and would take up too much room on the equipment. The oil properties will vary according to the individual needs of these devices. Non-smoking two-stroke oils are composed of esters or polyglycols. Environmental legislation for leisure marine applications, especially in Europe, encouraged the use of ester-based two cycle oil.

Properties

Most motor oils are made from a heavier, thicker petroleum hydrocarbon base stock derived from crude oil, with additives to improve certain properties. The bulk of a typical motor oil consists of hydrocarbons with between 18 and 34 carbon atoms per molecule. One of the most important properties of motor oil in maintaining a lubricating film between moving parts is its viscosity. The viscosity of a liquid can be thought of as its "thickness" or a measure of its resistance to flow. The viscosity must be high enough to maintain a lubricating film, but low enough that the oil can flow around the engine parts under all conditions. The viscosity index is a measure of how much the oil's viscosity changes as temperature changes. A higher viscosity index indicates the viscosity changes less with temperature than a lower viscosity index.

Motor oil must be able to flow adequately at the lowest temperature it is expected to experience in order to minimise metal to metal contact between moving parts upon starting up the engine. The *pour point* defined first this property of motor oil, as defined by ASTM D97 as "... an index of the lowest temperature of its utility ..." for a given application, but the "cold cranking simulator" and "Mini-Rotary Viscometre", ASTM D4684-08) are today the properties required in motor oil specs and define the SAE classifications.

Oil is largely composed of hydrocarbons which can burn if ignited. Still another important property of motor oil is its flash point, the lowest temperature at which the oil gives off vapours which can ignite. It is dangerous for the oil in a motor to ignite and burn, so a high flash point is desirable. At a petroleum refinery, fractional distillation separates a motor oil fraction from other crude oil fractions, removing the more volatile components, and therefore increasing the oil's flash point (reducing its tendency to burn).

Another manipulated property of motor oil is its Total Base Number (TBN), which is a measurement of the reserve alkalinity of an oil, meaning its ability to neutralise acids. The resulting quantity is determined as mg KOH/ (gram of lubricant). Analogously, Total Acid Number (TAN) is the measure of a lubricant's acidity. Other tests include zinc, phosphorus, or sulphur content, and testing for excessive foaming.

The NOACK volatility (ASTM D-5800) Test determines the physical evaporation loss of lubricants in high temperature service. A maximum of 15% evaporation loss is allowable to meet API SL and ILSAC GF-3 specifications. Some automotive OEM oil specifications require lower than 10%.

Grades

The Society of Automotive Engineers (SAE) has established a numerical code system for grading motor oils according to their viscosity characteristics. SAE viscosity gradings include the following, from low to high viscosity: 0, 5, 10, 15, 20, 25, 30, 40, 50 or 60. The numbers 0, 5, 10, 15 and 25 are suffixed with the letter W, designating their "winter" (not "weight") or cold-start viscosity, at lower temperature. The number 20 comes with or without a W, depending on whether it is being used to denote a cold or hot viscosity grade. The document SAE J300 defines the viscometrics related to these grades.

Kinematic viscosity is graded by measuring the time it takes for a standard amount of oil to flow through a standard orifice, at standard temperatures. The longer it takes, the higher the viscosity and thus higher SAE code. The SAE has a separate viscosity rating system for gear, axle, and manual transmission oils, SAE J306, which should not be confused with engine oil viscosity. The higher numbers of a gear oil (e.g., 75W-140) do not mean that it has higher viscosity than an engine oil.

Single-grade

A single-grade engine oil, as defined by SAE J300, cannot use a polymeric Viscosity Index Improver (also referred to as Viscosity Modifier) additive. SAE J300 has established eleven viscosity grades, of which six are considered Winter-grades and given a W designation. The 11 viscosity grades are 0W, 5W, 10W, 15W, 20W, 25W, 20, 30, 40, 50, and 60. These numbers are often referred to as the "weight" of a motor oil, and single-grade motor oils are often called "straight-weight" oils.

For single winter grade oils, the dynamic viscosity is measured at different cold temperatures, specified in J300 depending on the viscosity grade, in units of mPa·s, or the equivalent older non-SI units, centipoise (abbreviated cP), using two different test methods. They are the Cold Cranking Simulator and the Mini-Rotary Viscometre (ASTM D4684). Based on the coldest temperature the oil passes at, that oil is graded as SAE viscosity grade 0W, 5W, 10W, 15W, 20W, or 25W. The lower the viscosity grade, the lower the temperature the oil can pass. For example, if an oil passes at the specifications for 10W and 5W, but fails for 0W, then that oil must be labelled as an SAE 5W. That oil cannot be labelled as either 0W or 10W.

For single non-winter grade oils, the kinematic viscosity is measured at a temperature of 100 °C (212 °F) in units of mm^2/s (millimetre squared per second) or the equivalent older non-SI units, centistokes (abbreviated cSt). Based on the range of viscosity the oil falls in at that temperature, the oil is graded as SAE viscosity grade 20, 30, 40, 50, or 60. In addition, for SAE grades 20, 30, and 1000, a minimum viscosity measured at 150 °C (302 °F) and at a high-shear rate is also required. The higher the viscosity, the higher the SAE viscosity grade is. For some applications, such as when the temperature ranges in use are not very wide, single-grade motor oil is satisfactory; for example, lawn mower engines, industrial applications, and vintage or classic cars.

Multi-grade

The temperature range the oil is exposed to in most vehicles can be wide, ranging from cold temperatures in the winter before the vehicle is started up, to hot operating temperatures when the vehicle is fully warmed up in hot summer weather. A specific oil will have high viscosity when cold and a lower viscosity at the engine's operating temperature. The difference in viscosities for most single-grade oil is too large between the extremes of temperature. To bring the difference in viscosities closer together, special polymer additives called viscosity index improvers, or VIIs are added to the oil. These additives are used to make the oil a *multi-grade* motor oil, though it is possible to have a multi-grade oil without the use of VIIs. The idea is to cause the multi-grade oil to have the viscosity of the base grade when cold and the viscosity of the second grade when hot. This enables one type of oil to be used all year. In fact, when multi-grades were initially developed, they were frequently described as *all-season oil*. The viscosity of a multi-grade oil still varies logarithmically with temperature, but the slope representing the change is lessened. This slope representing the change with temperature depends on the nature and amount of the additives to the base oil.

The SAE designation for multi-grade oils includes two viscosity grades; for example, *10W-30* designates a common multi-grade oil. The two numbers used are individually defined by SAE J300 for single-grade oils. Therefore, an oil labelled as 10W-30 must pass the SAE J300 viscosity grade requirement for both 10W and 30, and all limitations placed on the viscosity grades (for example, a 10W-30 oil must fail the J300 requirements at 5W). Also, if an oil does not contain any VIIs, and can pass as a multi-grade, that oil can be labelled with either of the two SAE viscosity grades. For example, a very simple multi-grade oil that can be easily made with modern base oils without any VII is a 20W-20. This oil can be labelled as 20W-20, 20W, or 20. Note, if any VIIs are used however, then that oil cannot be labelled as a single grade.

The real-world ability of an oil to crank or pump when cold is potentially diminished soon after it is put into service. The motor oil grade and viscosity to be used in a given vehicle is specified by the manufacturer of the vehicle (although some modern European cars now have no viscosity requirement), but can vary from country to country when climatic or fuel efficiency constraints come into play.

Standards

American Petroleum Institute

The American Petroleum Institute (API) sets minimum for performance standards for lubricants. Motor oil is used for the lubrication, cooling, and cleaning of internal combustion engines. Motor oil may be composed of a lubricant base stock only in the case of non-detergent oil, or a lubricant base stock plus additives to improve the oil's detergency, extreme pressure performance, and ability to inhibit corrosion of engine parts. Lubricant base stocks are categorised into five groups by the API. Group I base stocks are composed of fractionally distilled petroleum which is further refined with solvent extraction processes to improve certain properties such as oxidation resistance and to remove wax. Group II base stocks are composed of fractionally distilled petroleum that has been hydrocracked to further refine and purify it. Group III base stocks have similar characteristics to Group II base stocks, except that Group III base stocks have higher viscosity indexes. Group III base stocks are produced by further hydrocracking of Group II base stocks, or of hydroisomerised slack wax, (a byproduct of the dewaxing process). Group IV base stock are polyalphaolefins (PAOs). Group V is a catch-all group for any base stock not described by Groups I to IV. Examples of group V base stocks includepolyol esters, polyalkylene glycols (PAG oils), and perfluoropolyalkylethers (PFPAEs). Groups I and II are commonly referred to as mineral oils, group III is typically referred to as synthetic (except in Germany and Japan, where they must not be called synthetic) and group IV is a synthetic oil. Group V base oils are so diverse that there is no catch-all description.

The API service classes have two general classifications: *S* for "service/spark ignition" (typical passenger cars and light trucks using gasoline engines), and *C* for "commercial/compression ignition" (typical dieselequipment). Engine oil which has been tested and meets the API standards may display the API Service Symbol (also known as the "Donut") with the service designation on containers sold to oil users. The API oil classification structure has eliminated specific support for wet-clutch motorcycle applications in their descriptors, and API SJ and newer oils are referred to be specific to automobile and light truck use. Accordingly, motorcycle oils are subject to their own unique standards.

The latest API service standard designation is SN for gasoline automobile and light-truck engines. The SN standard refers to a group of laboratory and engine tests, including the latest series for control of high-temperature deposits. Current API service categories include SN, SM, SL and SJ for gasoline engines. All previous service designations are obsolete, although motorcycle oils commonly still use the SF/SG standard.

All the current gasoline categories (including the obsolete SH), have placed limitations on the phosphorus content for certain SAE viscosity grades (the xW-20, xW-30) due to the chemical poisoning that phosphorus has on catalytic converters. Phosphorus is a key anti-wear component in motor oil and is usually found in motor oil in the form of zinc dithiophosphate. Each new API category has placed successively lower phosphorus and zinc limits, and thus has created a controversial issue of obsolescent oils needed for older engines, especially engines with sliding (flat/cleave) tappets. API, and ILSAC, which represents most of the worlds major automobile/engine manufactures, states API SM/ILSAC GF-4 is fully backwards compatible, and it is noted that one of the engine tests required for API SM, the Sequence IVA, is a sliding tappet design to test specifically for cam wear protection.

Not everyone is in agreement with backwards compatibility, and in addition, there are special situations, such as "performance" engines or fully race built engines, where the engine protection requirements are above and beyond API/ILSAC requirements. Because of this, there are speciality oils out in the market place with higher than API allowed phosphorus levels. Most engines built before 1985 have the flat/cleave bearing style systems of construction, which is sensitive to reducing zinc and phosphorus. Example; in API SG rated oils, this was at the 1200-1300 ppm level for zinc and phosphorus, where the current SM is under 600 ppm. This reduction in anti-wear chemicals in oil has caused premature failures of camshafts and other high pressure bearings in many older automobiles and has been blamed for pre-mature failure of the oil pump drive/cam position sensor gear that is meshed with camshaft gear in some modern engines.

There are six diesel engine service designations which are current: CJ-4, CI-4, CH-4, CG-4, CF-2, and CF. Some manufacturers continue to use obsolete designations such as CC for small or stationary diesel engines. In addition, API created a separated CI-4 PLUS designation in conjunction with CJ-4 and CI-4 for oils that meet certain extra requirements, and this marking is located in the lower portion of the API Service Symbol "Donut". It is possible for an oil to conform to both the gasoline and diesel standards. In fact, it is the norm for all diesel rated engine oils to carry the "corresponding" gasoline specification. For example, API CJ-4 will almost always list either SL or SM, API CI-4 with SL, API CH-4 with SJ, and so on.

ILSAC

The International Lubricant Standardisation and Approval Committee (ILSAC) also has standards for motor oil. Introduced in 2004, GF-4 applies to SAE 0W-20, 5W-20, 0W-30, 5W-30, and 10W-30 viscosity grade oils. In general, ILSAC works with API in creating the newest gasoline oil specification, with ILSAC adding an extra requirement of fuel economy testing to their specification. For GF-4, a Sequence VIB Fuel Economy Test (ASTM D6837) is required that is not required in API service category SM.

A key new test for GF-4, which is also required for API SM, is the Sequence IIIG, which involves running a 3.8 L (232 in^3), GM 3.8 L V-6 at 125 hp (93 kW), 3,600 rpm, and 150 °C (300 °F) oil temperature for 100 hours. These are much more severe conditions than any API-specified oil was designed for: cars which typically push their oil temperature consistently above 100 °C (212 °F) are most turbocharged engines, along with most engines of European or Japanese origin, particularly small capacity, high power output.

The IIIG test is about 50% more difficult than the previous IIIF test, used in GF-3 and API SL oils. Engine oils bearing the API starburst symbol since 2005 are ILSAC GF-4 compliant.

To help consumers recognise that an oil meets the ILSAC requirements, API developed a "starburst" certification mark.

A new set of specifications, GF-5, took effect in October 2010. The industry has one year to convert their oils to GF-5 and in September 2011, ILSAC will no longer offer licensing for GF-4.

ACEA

The ACEA (*Association des Constructeurs Européens d'Automobiles*) performance/quality classifications A3/A5 tests used in Europe are arguably more stringent than the API and ILSAC standards.

CEC (The Co-ordinating European Council) is the development body for fuel and lubricant testing in Europe and beyond, setting the standards via their European Industry groups; ACEA, ATIEL, ATC and CONCAWE. Lubrizol, a supplier of additives to nearly all motor oil companies, hosts a Relative Performance Tool which directly compares the manufacturer and industry specs. Differences in their performance is apparent in the form of interactive spider graphs, which both expert and novice can appreciate.

JASO

The Japanese Automotive Standards Organisation (JASO) has created their own set of performance and quality standards for petrol engines of Japanese origin. For four-stroke gasoline engines, the JASO T904 standard is used, and is particularly relevant to motorcycle engines. The JASO T904-MA and MA2 standards are designed to distinguish oils that are approved for wet clutch use, and the JASO T904-MB standard is not suitable for wet clutch use.

For two-stroke gasoline engines, the JASO M345 (FA, FB, FC) standard is used, and this refers particularly to low ash, lubricity, detergency, low smoke and exhaust blocking. These standards, especially JASO-MA and JASO-FC, are designed to address oil-requirement issues not addressed by the API service categories.

OEM Standards Divergence

By the early 1990s, many of the European original equipment manufacturer (OEM) car manufacturers resigned on the lacklustre direction of the American API oil standards as it did not perform to the needs of a motor oil to be used in their motors and seriously lagged in development of the previous generations. As a result many leading European motor manufacturers created and developed their own "OEM" oil standards which were no longer directly compatible with the plain API. (Note that the ACEA class of standards is co-developped with all European engine makers to better suit the legislative and technical needs, thus ACEA specification on the back label completely or nearly conforms to many OEM specifications.) In recent years similar happened in the North American diesel engine market in the high performance segment, with names such as Caterpillar, John Deere, Mack, Cummins, Ford appearing on the back of the oil cans in lists of certifications. It is irony that the standard "API C" means "Commercial", and yet it failed to fullfill the needs of the main commercial engine producers.

Some of the of widely used OEM standards are the VW500.**, VW505.** series from Volkswagen Group, and the MB228.* and

MB229.** from Mercedes-Benz. Other European OEM standards are from General Motors (dexos), the Ford "WSS" standards, BMW Special Oils and BMW Longlife standards, Porsche, and the PSA Group of Peugeot and Citroën. Prior to the development of the dexos standard, General Motors used the 4718M standard that is used for high-performance engines, a standard that is used in North America for selected North American performance engines, with a "Use Mobil 1 only" sticker was usually placed on those cars.

More recently, "extended drain", "BMW longlife" and similar oils have arisen, whereby, taking Volkswagen Group vehicles, a petrol engine can go up to 2 years or 30,000 km (~18,600 mi), and a diesel engine can go up to 2 years or 50,000 km (~31,000 mi) — before requiring an oil change. Volkswagen (504.00), BMW, GM, Mercedes and PSA all have their own similar longlife oil standards. (In case of MB certified oils, the standard applies to oils used in trucks and personal cars alike, so every Mercedes engine was expected to use the same oil, while other small car makers were satisfied with the generic — and thus least performing lubricants available and caught up recently when the legislation from the European Council mandated them to improve fuel consumption and improve emissions.)

This should be not surprising as Mercedes was among the first to differentiate oils according to longevity (1980s to 1990). In personal vehicles is a general oil change interval given by oil specification: 228.1 – 15000 km, 228.3 – 30000 km, 228.5 – 45000 km. (Similar rule applies to the MB 229.xx) Oil certified for the longest change interval also had the best antioxidative properties and stability. Certain BP Vanellus oil certified for MB228.5 standard had sulphated ash content around 2%, thus providing superior piston ring protection as a side-effect. Such oils were originally marketed for heavy trucking use (100000 miles change interval) and other "long life" oils are likely to be of similar grade. The North American habit of having oil changed in the engine every 3000 miles has its roots if past far ago, when the API SC, CB oils were the norm. Those had reserve alkalinity and buffering ability only sourced from the bulk mass of the fresh base stock and offered very, very little in terms of surface protection in corrosion or mechanical resistance. With better lubricants in the beginning of the 1980s, in Europe longer service intervals became the norm, with 10000 km in standard car use as the typical value in the 1990. Many service technicians still recommend 3000 or 5000 miles service intervals in the conservative North American market, as it suits them as a source of revenue and also there is less of a need to provide top quality lubricants.

Another trend of today represent midSAP (sulphated ash <0,8 wt.-%) and lowSAP (sulphated ash <0,5 wt.-%) engine oil. The ACEA specifications C1 to C4 reflect the midSAP and lowSAP needs of automotive OEMs. Reason for this is that the small loss of the oil during engine life — that ends burned, exhaust valve stem cleanliness is improved and much less gypsum ends up plugging the catalyst and/or particulate filter. This improves emissions as the emission regulating system has longer service life then. It is much more remarkable in a trailer truck which can easily make 200000 to 400000 km every year.

Furthermore, European ACEA standards require that during the long drain intervals of 30.000 km and up HTHS viscosity is maintained (High Temperature, High Shear) at around 3.5 cP (3.5 mPa·s). Required minimum HTHS viscosity is given by SAE oil grade. SAE 40 needs at least 3.5cSt, SAE 30 — 2.9cSt, etc. This is important when considering oil change in an engine. A too low HTHS viscosity and protection of piston rings and journal bearings may be compromised. Thus engines requiring SAE 0W-20 oil do so because their operating temperatures are far lower than those using SAE-40 viscosity oil. As in API, no system of marking that the product exceeds this critical specification by a significant amount is in place.

Because of the need for motor oils with unique qualities, many modern cars for the European market will demand a specific OEM-only oil standard. As a result, it may make no reference at all to ACEA or API standards. This is case of VW pumpe-düse diesel engines, as the manufacturer cannot guarantee longevity and reliability of a certain engine components without adherence to the specification. While it may be confusing that the standard may not specify SAE viscosity, it is not the important parameter. Wear protection and HTHS viscosity are important parameters and are not specified in the SAE viscosity standard. Additionally, API tests are performed on engines subjected to far lighter loads and shears. The reason for current development of new OEM standards is that in the 1970s to 1980s when the SAE and API refused to develop standards for characterising oils by their HTHS viscosity or by their lubricating properties, because 'some products would be looking bad even if they were completely OK'

Quote from ASTM report on the matter: "The rapid growth of non-Newtonian multigraded oils has rendered kinematic viscosity as a nearly useless parameter for characterising "real" viscosity in critical zones of an engine.

... There are those who are disappointed that the twelve-year effort has not resulted in a redefinition of the SAE J300 Engine Oil Viscosity

Classification document so as to express high-temperature viscosity of the various grades ...

... In the view of this writer, this redefinition did not occur because the automotive lubricant market knows of no field failures unambiguously attributable to insufficient HTHS oil viscosity."

Since low and high quality oils conformed to the same standard which did not mandate critical parameters, engine manufacturers were FORCED to develop their own standards and tests, as the lubricant providers did NOT manufacture lubricants with guaranteed minimal lubricity under real world stress conditions at the time of the development of engines conforming to new legislation. As new lubricant standards were only always introduced after lengthy proceedings to arrive barely in time with the new generation of engines, users were always left in the dark when comparing various oil brands and products which all conformed to the same maximum specification, even if particular products could perform far better than others.

All this could be prevented 40 years ago when inclusion of HTHS standards was demanded by lubrication experts into any kind of standard.

Thus user today who wants to top up or change engine oil needs to pay close attention to the list of certificates on the oil label (on the back side), and gain understanding of the specific manufacturer designation meaning, which number signifies petrol engine, which diesel engine. Which number marks suitability for a turbocharged engine, etc.

Other Additives

In addition to the viscosity index improvers, motor oil manufacturers often include other additives such as detergents and dispersants to help keep the engine clean by minimising sludge buildup, corrosion inhibitors, and alkaline additives to neutralise acidic oxidation products of the oil. Most commercial oils have a minimal amount of zinc dialkyldithiophosphate as an anti-wear additive to protect contacting metal surfaces with zinc and other compounds in case of metal to metal contact. The quantity of zinc dialkyldithiophosphate is limited to minimise adverse effect on catalytic converters. Another aspect for after-treatment devices is the deposition of oil ash, which increases the exhaust back pressure and reduces fuel economy over time. The so-called "chemical box" limits today the concentrations of sulphur, ash and phosphorus (SAP).

There are other additives available commercially which can be added to the oil by the user for purported additional benefit. Some of these additives include:

- EP additives, like zinc dialkyldithiophosphate (ZDDP) additives and sulphonates, preferably calcium sulphonates, are available to consumers for additional protection under extreme-pressure conditions or in heavy duty performance situations. Calcium sulphonates additives are also added to protect motor oil from oxidative breakdown and to prevent the formation of sludge and varnish deposits. Both were the main basis of additive packages used by lubricant manufacturers up until 1990s when the need for ashless addtitives arose. Main advantage was very low price and wide availability (sulphonates were originally waste byproducts). Currently there are ashless oil lubricants without these additives, which can only fulfill the qualities of the previous generation with more expensive basestock and more expensive organic or organometallic additive compounds. Some new oils are not formulated to provide the level of protection of previous generations to save manufacturing costs. Lately API specifications reflect that
- Some molybdenum disulphide containing additives to lubricating oils are claimed to reduce friction, bond to metal, or have anti-wear properties. MoS_2 particles can be shear-welded on steel surface and some engine components were even treated with MoS_2 layer during manufacture, namely liners in engines. (Trabant for example). They were used in World War II in flight engines and became commercial after World War II until the 1990s. They were commercialised in the 1970s (ELF ANTAR Molygraphite) and are today still available. Main disadvantage of molybdenum disulphide is anthracite black colour, so oil treated with it is hard to distinguish from a soot filled engine oil with metal shavings from spun crankshaft bearing.
- In the 1980s and 1990s, additives with suspended PTFE particles were available, e.g., "Slick50", to consumers to increase motor oil's ability to coat and protect metal surfaces. There is controversy as to the actual effectiveness of these products, as they can coagulate and clog the oil filters. It is supposed to work under boundary lubricating conditions, which good engine designs tend to avoid anyway. Also, Teflon alone has little to no ability to firmly stick on a sheared surface, unlike molybdenum disulphide, for example.

- Various other extreme-pressure additives and antiwear additives.
- Many patents proposed use perfluoropolymers to reduce friction between metal parts, such as PTFE (Teflon), or micronised PTFE. However, the application obstacle of PTFE is insolubility in lubricant oils. Their application is questionable and depends mainly on the engine design — one that can not maintain reasonable lubricating conditions might benefit, while properly designed engine with oil film thick enough would not see any difference. Other invalid claim about PTFE is the friction factor as it depends on material hardness. PTFE is a very soft material, thus its friction coefficient becomes worse than that of hardened steel-to-steel mating surfaces under common loads. PTFE is used in composition of sliding bearings where it improves lubrication under relatively light load until the oil pressure builds up to full hydrodynamic lubricating conditions.

Chapter 8

Oil Reserves

The total estimated amount of oil in an oil reservoir, including both producible and non-producible oil, is called *oil in place*. However, because of reservoir characteristics and limitations in petroleum extraction technologies, only a fraction of this oil can be brought to the surface, and it is only this producible fraction that is considered to be *reserves*. The ratio of producible oil reserves to total oil in place for a given field is often referred to as the *recovery factor*. Recovery factors vary greatly among oil fields. The recovery factor of any particular field may change over time based on operating history and in response to changes in technology and economics. The recovery factor may also rise over time if additional investment is made in enhanced oil recovery techniques such as gas injection, surfactants injection, water-flooding, or microbial enhanced oil recovery.

Based on data from OPEC at the beginning of 2011 the highest proved oil reserves including non-conventional oil deposits are in Venezuela (20% of global reserves), Saudi Arabia (18% of global reserves), Iran (9%).

Because the geology of the subsurface cannot be examined directly, indirect techniques must be used to estimate the size and recoverability of the resource. While new technologies have increased the accuracy of these techniques, significant uncertainties still remain. In general, most early estimates of the reserves of an oil field are conservative and tend to grow with time. This phenomenon is called *reserves growth.*

Many oil-producing nations do not reveal their reservoir engineering field data and instead provide unaudited claims for their oil reserves. The numbers disclosed by some national governments are suspected of being manipulated for political reasons.

Classifications

All reserve estimates involve uncertainty, depending on the amount of reliable geologic and engineering data available and the interpretation of those data. The relative degree of uncertainty can be expressed by dividing reserves into two principal classifications—"proven" (or "proved") and "unproven" (or "unproved"). Unproven reserves can further be divided into two subcategories—"probable" and "possible"—to indicate the relative degree of uncertainty about their existence. The most commonly accepted definitions of these are based on those approved by the Society of Petroleum Engineers (SPE) and the World Petroleum Council (WPC) in 1997.

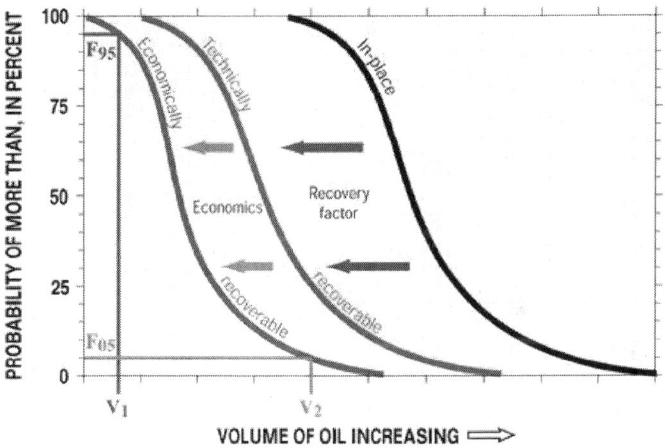

Figure: *Schematic graph illustrating petroleum volumes and probabilities. Curves represent categories of oil in assessment. There is a 95% chance (i.e., probability, F95) of at least volume V1 of economically recoverable oil, and there is a 5-percent chance (F05) of at least volume V2 of economically recoverable oil.*

Proven Reserves

Proven reserves are those reserves claimed to have a *reasonable certainty* (normally at least 90% confidence) of being recoverable under existing economic and political conditions, with existing technology. Industry specialists refer to this as P90 (i.e., having a 90% certainty of being produced). Proven reserves are also known in the industry as 1P.

Proven reserves are further subdivided into "proven developed" (PD) and "proven undeveloped" (PUD). PD reserves are reserves that can be produced with existing wells and perforations, or from additional reservoirs where minimal additional investment (operating expense) is required. PUD reserves require additional capital investment (e.g., drilling new wells) to bring the oil to the surface.

Until December 2009 "1P" proven reserves were the only type the U.S. Securities and Exchange Commission allowed oil companies to report to investors. Companies listed on U.S. stock exchanges must substantiate their claims, but many governments and national oil companies do not disclose verifying data to support their claims. Since January 2010 the SEC now allows companies to also provide additional optional information declaring "2P" (both proven and probable) and "3P" (proven + probable + possible) provided the evaluation is verified by qualified third party consultants, though many companies choose to use 2P and 3P estimates only for internal purposes.

Unproven Reserves

Unproven reserves are based on geological and/or engineering data similar to that used in estimates of proven reserves, but technical, contractual, or regulatory uncertainties preclude such reserves being classified as proven. Unproven reserves may be used internally by oil companies and government agencies for future planning purposes but are not routinely compiled. They are sub-classified as *probable* and *possible*. Probable reserves are attributed to known accumulations and claim a 50% confidence level of recovery. Industry specialists refer to them as "P50" (i.e., having a 50% certainty of being produced). These reserves are also referred to in the industry as "2P" (proven plus probable).

Possible reserves are attributed to known accumulations that have a less likely chance of being recovered than probable reserves. This term is often used for reserves which are claimed to have at least a 10% certainty of being produced ("P10").

Reasons for classifying reserves as possible include varying interpretations of geology, reserves not producible at commercial rates, uncertainty due to reserve in fill and projected reserves based on future recovery methods. They are referred to in the industry as "3P" (proven plus probable plus possible).

Fraud and incompetence: In the early 1980s the Penn Square Bank of Oklahoma engaged in wide scale lending to Oil and Gas groups with reserves used as collateral for the loans. Sometimes the bank officers ignored the engineering advice about reserve size and value in order to book the loan profit. The bank resold many of the loans as 'loan participations' to larger banks like Continental Illinois, Seattle First, Michigan National, and Chase Manhattan, whose officers also sometimes ignored engineering advice. When the oil market busted, many of these loans were called, and the true nature of the reserves used as loan collateral was discovered by Federal Deposit Insurance Corporation government investigators and crors. Penn Square went bust, and Continental Illinois was the first Too Big to Fail bailout of the modern era.

Strategic Petroleum Reserves

Many countries maintain government-controlled oil reserves for both economic and national security reasons. According to the United States Energy Information Administration, approximately 4.1 billion barrels (650,000,000 m^3) of oil are held in strategic reserves, of which 1.4 billion is government-controlled (m3=cubic metres). These reserves are generally not counted when computing a nation's oil reserves.

Resources

A more sophisticated system of evaluating petroleum accumulations was adopted in 2007 by the Society of Petroleum Engineers (SPE), World Petroleum Council (WPC), American Association of Petroleum Geologists (AAPG), and Society of Petroleum Evaluation Engineers (SPEE). It incorporates the 1997 definitions for reserves, but adds categories for *contingent resources* and *prospective resources*.

Contingent resources are those quantities of petroleum estimated, as of a given date, to be potentially recoverable from *known* accumulations, but the applied project(s) are not yet considered mature enough for commercial development due to one or more contingencies. Contingent resources may include, for example, projects for which there are currently no viable markets, or where commercial recovery is dependent on technology under development, or where evaluation of the accumulation is insufficient to clearly assess commerciality. Prospective resources are those quantities of petroleum estimated, as of a given date, to be potentially recoverable from *undiscovered* accumulations by application of future development projects. Prospective resources have both an associated chance of discovery and a chance of development.

The United States Geological Survey uses the terms *technically* and *economically* recoverable resources when making its petroleum resource assessments. Technically recoverable resources represent that proportion of assessed in-place petroleum that may be recoverable using current recovery technology, without regard to cost. Economically recoverable resources are technically recoverable petroleum for which the costs of discovery, development, production, and transport, including a return to capital, can be recovered at a given market price.

"Unconventional resources" exist in petroleum accumulations that are pervasive throughout a large area. Examples include extra heavy oil, natural bitumen, and oil shaledeposits. Unlike "conventional resources", in which the petroleum is recovered through wellbores and typically requires minimal processing prior to sale, unconventional resources require specialised extraction technology to produce. For example, steam and/or solvents are used to mobilise bitumen for in-situ recovery. Moreover, the extracted petroleum may require significant processing prior to sale (e.g., bitumen upgraders). The total amount of unconventional oil resources in the world considerably exceeds the amount of conventional oil reserves, but are much more difficult and expensive to develop.

Estimation Techniques

The amount of oil in a subsurface reservoir is called *oil in place* (OIP). Only a fraction of this oil can be recovered from a reservoir. This fraction is called the *recovery factor*. The portion that can be recovered is considered to be a reserve. The portion that is not recoverable is not included unless and until methods are implemented to produce it.

There are a number of different methods of calculating oil reserves. These methods can be grouped into three general categories: "volumetric", "material balance", and "production performance". Each method has its advantages and drawbacks.

Volumetric Method

Volumetric methods attempt to determine the amount of oil in place by using the size of the reservoir as well as the physical properties of its rocks and fluids. Then a recovery factor is assumed, using assumptions from fields with similar characteristics. OIP is multiplied by the recovery factor to arrive at a reserve number. Current recovery factors for oil fields around the world typically range between 10 and 60 percent; some are over 80 percent. The wide variance is due largely to the diversity of fluid and reservoir characteristics for different deposits. The method is most useful early in the life of the reservoir, before significant production has occurred.

Materials Balance Method

The *materials balance method* for an oil field uses an equation that relates the volume of oil, water and gas that has been produced from a reservoir and the change in reservoir pressure to calculate the remaining oil. It assumes that, as fluids from the reservoir are produced, there will be a change in the reservoir pressure that depends on the remaining volume of oil and gas.

The method requires extensive pressure-volume-temperature analysis and an accurate pressure history of the field. It requires some production to occur (typically 5% to 10% of ultimate recovery), unless reliable pressure history can be used from a field with similar rock and fluid characteristics.

Production Decline Curve Method

The *decline curve method* uses production data to fit a decline curve and estimate future oil production. The three most common forms of decline curves are exponential, hyperbolic, and harmonic. It is assumed that the production will decline on a reasonably smooth curve, and so allowances must be made for wells shut in and production restrictions. The curve can be expressed mathematically or plotted on a graph to estimate future production. It has the advantage of (implicitly) including all reservoir characteristics. It requires a sufficient history to establish a statistically significant trend, ideally when production is not curtailed by regulatory or other artificial conditions.

Reserves Growth

Experience shows that initial estimates of the size of newly discovered oil fields are usually too low. As years pass, successive estimates of the ultimate recovery of fields tend to increase. The term *reserve growth* refers to the typical increases in estimated ultimate recovery that occur as oil fields are developed and produced.

Estimated Reserves by Country

OPEC Countries

There are doubts about the reliability of official OPEC reserves estimates, which are not provided with any form of lubrication or verification that meet external reporting standards.

Since a system of country production quotas was introduced in the 1980s, partly based on reserves levels, there have been dramatic increases in reported reserves among OPEC producers. In 1983, Kuwait increased its proven reserves from 67 Gbbl (10.7×10^9 m^3) to 92 Gbbl (14.6×10^9 m^3). In 1985–86, the UAE almost tripled its reserves from 33 Gbbl (5.2×10^9 m^3) to 97 Gbbl (15.4×10^9 m^3). Saudi Arabia raised its reported reserve number in 1988 by 50%. In 2001–02, Iran raised its proven reserves by some 30% to 130 Gbbl (21×10^9 m^3), which advanced it to second place in reserves and ahead of Iraq. Iran denied accusations of a political motive behind the readjustment, attributing the increase instead to a combination of new discoveries and improved recovery. No details were offered of how any of the upgrades were arrived at.

The following table illustrates these rises.

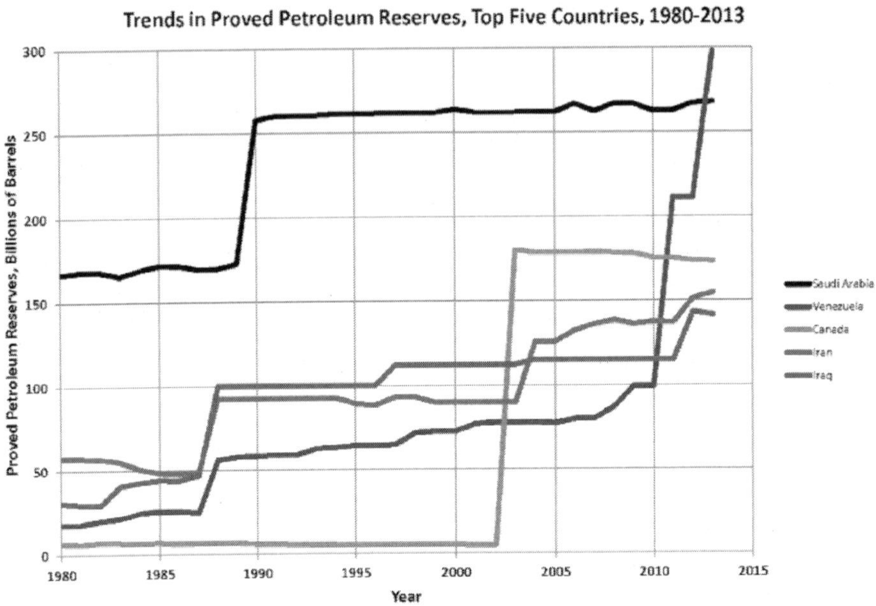

Figure: *Trends in proved oil reserves in top five countries, 1980-2013 (date from US Energy Information Administration)*

The sudden revisions in OPEC reserves, totalling nearly 300 bn barrels, have been much debated. Some of it is defended partly by the shift in ownership of reserves away from international oil companies, some of whom were obliged to report reserves under conservative US Securities and Exchange Commission rules. The most prominent explanation of the revisions is prompted by a change in OPEC rules which set production quotas (partly) on reserves. In any event, the revisions in official data had little to do with the actual discovery of new reserves.

Total reserves in many OPEC countries hardly changed in the 1990s. Official reserves in Kuwait, for example, were unchanged at 96.5 Gbbl (15.34×10^{9} m^3) (including its share of the Neutral Zone) from 1991 to 2002, even though the country produced more than 8 Gbbl (1.3×10^{9} m^3) and did not make any important new discoveries during that period. The case of Saudi Arabia is also striking, with proven reserves estimated at between 260 and 264 billion barrels (4.20×10^{10} m^3) in the past 18 years, a variation of less than 2%, while extracting approximately 60 billion barrels (9.5×10^{9} m^3) during this period.

Sadad al-Huseini, former head of exploration and production at Saudi Aramco, estimates 300 Gbbl (48×10^{9} m^3) of the world's 1,200 Gbbl (190×10^{9} m^3) of proven reserves should be recategorised as speculative resources, though he did not specify which countries had inflated their reserves. Dr. Ali Samsam Bakhtiari, a former senior expert of the National Iranian Oil Company, has estimated that Iran, Iraq, Kuwait, Saudi Arabia and the United Arab Emirates have overstated reserves by a combined 320–390bn barrels and has said, "As for Iran, the usually accepted official 132 billion barrels (2.10×10^{10} m^3) is almost one hundred billion over any realistic assay." *Petroleum Intelligence Weekly* reported that official confidential Kuwaiti documents estimate reserves of Kuwait were only 48 billion barrels (7.6×10^9 m^3), of which half were proven and half were possible. The combined value of proven and possible is half of the official public estimate of proven reserves.

In July 2011, OPEC's Annual Statistical Review showed Venezuela's reserves to be larger than Saudi Arabia's.

Prospective Resources

Arctic Prospective Resources

A 2008 United States Geological Survey estimates that areas north of the Arctic Circle have 90 billion barrels (1.4×10^{10} m^3) of undiscovered, technically recoverable oil and 44 billion barrels (7.0×10^9 m^3) of natural gas liquids in 25 geologically defined areas thought to have potential for petroleum. This represented 13% of the expected undiscovered oil in the world. Of the estimated totals, more than half of the undiscovered oil resources were estimated to occur in just three geologic provinces—Arctic Alaska, the Amerasia Basin, and the East Greenland Rift Basins. More than 70% of the mean undiscovered oil resources was estimated to occur in five provinces: Arctic Alaska, Amerasia Basin, East Greenland Rift Basins, East Barents Basins, and West Greenland–East Canada. It was further estimated that approximately 84% of the oil and gas would occur offshore. The USGS did not consider economic factors such as the effects of permanent sea ice or oceanic water depth in its assessment of undiscovered oil and gas resources. This assessment was lower than a 2000 survey, which had included lands south of the Arctic Circle.

Miscellaneous Prospective Resources

In October 2009, the USGS updated the Orinoco tar sands (Venezuela) value to 513 billion barrels (8.16×10^{10} m^3).

The Exclusive Economic Zone is a delineated offshore area, mostly to the west and north of Cuba, which, under international agreements, is owned by Cuba. This 112,000-square-kilometre zone has been divided into 59 exploration blocks. A 2004 joint partnership between a Spanish oil company and Cuba's state oil company (CUPET) estimated Cuba's off-shore reserves to be able to produce ultimately between 4.6 and 9.3 billion barrels of crude oil.

The US Geological Survey (USGS) estimates that Cuba has resources up to 9 billion barrels (1.4×10^9 m^3) of oil. In October 2008, the Cuban government announced that it had discovered oil basins that would double its total oil resources to 20 billion barrels (3.2×10^9 m^3). Note that, consistent with the definitions of reserves and resources as given in the above sections, at this stage (June 2009), given no commercial discoveries to date, all recoverable oil estimates in the Exclusive Economic Zone (EEZ) are prospective resources estimates and not reserves estimates.

Global Strategic Petroleum Reserves

Global strategic petroleum reserves ("GSPR") refer to crude oil inventories (or stockpiles) held by the government of a particular country, as well as private industry, for the purpose of providing economic and national security during an energy crisis. According to the United States Energy Information Administration, approximately 4.1 billion barrels (650,000,000 m^3) of oil are held in strategic reserves, of which 1.4 billion is government-controlled. The remainder is held by private industry. At the moment the US Strategic Petroleum Reserve is one of the largest strategic reserves, with much of the remainder held by the other 26 members of the International Energy Agency. Other non-IEA countries have begun creating their own strategic petroleum reserves, with China being the largest of these new reserves. Since current consumption levels are neighbouring 0.1 billion barrels (16,000,000 m^3) per day, in the case of a dramatic worldwide drop in oil field output as suggested by some peak oil analysts, the strategic petroleum reserves are unlikely to last for more than a few months.

International Energy Agency reserves

According to a March 2001 agreement, all 28 members of the International Energy Agency must have a strategic petroleum reserve equal to 90 days of prior year's net oil imports for their respective country. Only net-exporter members of the IEA are exempt from the reserve requirement. The exempt countries are Canada, Denmark, Norway, and the United Kingdom. However, Denmark and the UK have both recently created strategic reserves due to their requirements as European Union members.

Forward Commercial Storage Agreements

To allow oil-exporting countries increased flexibility in their production quotas, there has been an increased movement towards forward commercial storage agreements. These agreements allow petroleum to be stored at an oil-importing country, however the reserves are technically under the control of the oil-exporting country. Oil importing countries benefit from the close access to the commercial reserves, while reducing the costs of access.

Emergency Oil Sharing Agreements

In addition to maintaining a domestic stockpile of petroleum, several countries also have agreements to share their stockpiles in the event of an emergency.

The Japan, New Zealand and South Korea Agreement

In mid-2007 Japan announced a program to share its strategic reserve with other countries in its region. Negotiations are currently underway with New Zealand on an emergency oil-sharing program whereby Japan would make available for purchase its strategic reserves. In an emergency New Zealand would pay the market price plus negotiated option fees for the amount of oil previously held for them by Japan. South Korea and Japan have also agreed to share their oil reserves in case of an emergency.

United States and Israel Agreement

According to the 1975 Second Sinai withdrawal document signed by the United States and Israel, in an emergency the U.S. is obligated to make oil available for sale to Israel for up to 5 years.

The France, Germany, and Italy Agreement

France, Germany and Italy have an oil-sharing agreement to allow each other to purchase their strategic reserves in the event of an emergency.

Africa

South Africa has an SPR. It is managed by PetroSA and the primary facility is the Saldanha Bay oil storage facility, which is a major transit point for oil shipping. Saldanha Bay's six in-ground concrete storage tanks give the facility a storage capacity of 45,000,000 barrels (7,200,000 m^3).

Malawi is considering creating a 21-day reserve of fuel, which is an expansion from the current five day reserve. The government has begun planning for storage facilities in the provinces of Chipoka and Mchinji as well as Kamuzu International Airport.

Kenya is setting up a Strategic Fuel Reserve, similar to that of cereals. The strategic stocks would be procured by the National Oil Corporation of Kenya and stored by the Kenya Pipeline Company Limited.

Asia

China: In 2007 China announced an expansion of their crude reserves into a two part system. Chinese reserves would consist of a government-controlled strategic reserve complemented by mandated commercial reserves. The government-controlled reserves are being completed in three phases. Phase one consisted of a 101,900,000 barrels (16,200,000 m^3) reserve, mostly completed by the end of 2008. The second phase of the government-controlled reserves with an additional 170,000,000 barrels (27,000,000 m^3) will be completed by 2011. Recently, Zhang Guobao the head of the National Energy Administration also stated that there will be a third phase that will expand reserves by 204,000,000 barrels (32,400,000 m^3) with the goal of increasing China's SPR to 90 days of supply by 2020.

The planned state reserves of 475,900,000 barrels (75,660,000 m^3) plus the planned enterprise reserves of 209,440,000 barrels (33,298,000 m^3) will provide around 90 days of consumption or a total of 684,340,000 barrels (108,801,000 m^3).

India: India has begun the development of a strategic crude oil reserve sized at 37,400,000 barrels (5,950,000 m^3), enough for two weeks of consumption. Petroleum stocks have been transferred from the Indian Oil Corporation (IndianOil) to the Oil Industry Development Board (OIDB). The OIDB then created the Indian Strategic Petroleum Reserves Ltd (ISPRL) to serve as the controlling government agency for the strategic reserve.

The facilities are:

- Mangalore, State of Karnataka. Capacity of 11.22 million barrels (1,784,000 m^3).
- Padur village, Udipi in the state of Karnataka. Capacity of 18.7 million barrels (2,970,000 m^3).
- Visakhapatnam, State of Andhra Pradesh. Capacity of 7.48 million barrels (1,189,000 m^3).

On 21st December 2011, it was announced that India planned to greatly augment their crude reserve capacity to 132 million barrels by 2020, a senior oil ministry official stated.

- State controlled reserves of petroleum at eleven different locations totalling 324,000,000 barrels (51,500,000 m^3). *All numbers, unless otherwise cited, come from p. 177 of this document: 1*
- Tomakomai Eastern Oil Reserve Storage Base, 55 storage tanks, total capacity 34 million barrels (5,400,000 m^3).

- Mutsu-Ogawara Storage Base, 53 storage tanks, total capacity 31 million barrels (4,900,000 m^3).
- Kuji Storage Base, 3 storage tanks, total capacity 10.5 million barrels (1,670,000 m^3).
- Akita Storage Base, 15 storage tanks, total capacity 23.4 million barrels (3,720,000 m^3).
- Fukui Storage Base, 27 storage tanks, total capacity 17.9 million barrels (2,850,000 m^3).
- Kikuma Underground Petroleum Storage Facility, 8 storage tanks, total capacity 8.9 million barrels (1,410,000 m^3).
- Shirashima Storage Facility, 8 tankers (4,400,000 barrels (700,000 m^3) each), total capacity 35.2 million barrels (5,600,000 m^3).
- Kamigotou Storage Base, 7 storage tanks, total capacity 21.45 million barrels (3,410,000 m^3).
- Kushikino Storage Base, 3 storage tanks, total capacity 10.5 million barrels (1,670,000 m^3).
- Shibushi Storage Base, 40 storage tanks, total capacity 27.6 million barrels (4,390,000 m^3).
- Kagoshima former Nippon Oil facility, 4,000,000 barrels (640,000 m^3). This is a forward commercial storage facility with Abu Dhabi.
- Privately held reserves of petroleum held "in accordance with the Petroleum Stockpiling Law" of 129,000,000 barrels (20,500,000 m^3).
- Privately held reserves of petroleum products for another 130,000,000 barrels (21,000,000 m^3).

The state stockpile and the privately held stockpiles total about 583,000,000 barrels (92,700,000 m^3). The Japanese SPR is run by the Japan Oil, Gas and Metals National Corporation

South Korea: As part of the government's energy security efforts, South Korea holds strategic oil reserves to protect against oil supply disruptions. The country's strategic oil reserve program is managed by the Korea National Oil Corporation, which reports that its system has the capacity to store 116,000,000 barrels (18,400,000 m^3) of oil. As of April 2007, KNOC held 76,000,000 barrels (12,100,000 m^3) of oil in its strategic stockpiles, 64,000,000 barrels (10,200,000 m^3) of crude oil and 12,000,000 barrels (1,900,000 m^3) of petroleum products.

This total amounts to approximately 34 days of net import cover, according to 2006 estimates of demand. KNOC has plans to expand the country's strategic storage capacity from 116,000,000 barrels (18,400,000 m^3) to 146,000,000 barrels (23,200,000 m^3) by 2009, and to fill the emergency reserves to 141,000,000 barrels (22,400,000 m^3) by 2010.

Others

The Philippines has begun plans for a National Petroleum Strategic Reserve by 2010 with an approximate size of 30,000,000 barrels (4,800,000 m^3). Russia has begun plans for a strategic petroleum reserve. Analysts estimate the size of the Russian SPR would be around 78,000,000 barrels (12,400,000 m^3). Singapore has an SPR composed of 31.8 million barrels (5,060,000 m^3) of crude oil with an additional 64.5 million barrels (10,250,000 m^3) of oil products for a total of 96,300,000 barrels (15,310,000 m^3).

Taiwan has an SPR with a 1999 reported size of 13,000,000 barrels (2,100,000 m^3). Taiwan's refiners (Kaohsiung 270,000 bbl/d (43,000 m^3/d); Ta-Lin 300,000 bbl/d (48,000 m^3/d); Tao-Yuan 200,000 bbl/d (32,000 m^3/d); Mailiao 150,000 bbl/d) are also required to store at least 30 days of petroleum stocks. As of 2005, these mandated commercial reserves total 27,600,000 barrels (4,390,000 m^3) of strategic petroleum stocks.

Thailand has increased the size of its SPR from 60 days to 70 days of consumption in 2006.

Pakistan has begun plans for a 20 day emergency reserve.

Europe

European Union: In the European Union, according to Council Directive 68/414/EEC of 20 December 1968, all 27 members must have a strategic petroleum reserve within the territory of the E.U. equal to at least 90 days average daily internal consumption.

The Czech Republic has a four tank SPR facility in Nelahozeves run by the company CR Mero. The Czech SPR is equal to 100 days of consumption or 20,300,000 barrels (3,230,000 m^3).

Denmark has a reserve of 81 days of consumption, equal to about 1,4 million tonnes of oil products. Not counting reserves held by the military defence.

Finland has an SPR with an approximate size of 62,400,000 barrels (9,920,000 m^3).

France has an SPR with an approximate size of 65,000,000 barrels (10,300,000 m^3). As of 2000 jet fuel stocks were required for at least 55 days of consumption, with half of those stocks controlled by the *Société Anonyme de Gestion des Stocks de Sécurité* (SAGESS) and the other half controlled by producers.

Germany created the Federal Oil Reserve in 1970, stored in the Etzel salt caverns near Wilhelmshaven in northern Germany, with an initial size of 70 million barrels (11,000,000 m^3). The current German Federal Oil Reserve and the Erdölbevorratungsverband (EBV) (the German stockholding company) mandates that refiners must keep 90 days of stock on hand, giving Germany an approximate reserve size of 250,000,000 barrels (40,000,000 m^3) as of 1997. The German SPR is the largest in Europe.

Hungary has an SPR with approximately 90 days of consumption or 11,880,000 barrels (1,889,000 m^3).

Ireland has approximately 31 days of oil stocks in Ireland and another 9 days of oil stocks held in fellow EU members states. Additionally, they have stock tickets (contracts with a 3rd party where the government has the option to purchase in the event of an emergency) and stocks held by large industry or large consumers. On average Ireland has approximately 100 days of oil available.

Poland has an SPR with approximately 70 days of consumption. Another facility holding 20 additional days of consumption is scheduled to be completed in 2008. Poland also requires oil companies to maintain reserves sufficient for 73 days of production.

Portugal has an SPR with an approximate size of 22,440,000 barrels (3,568,000 m^3).

Slovakia has an SPR with an approximate size of 748,000 barrels (118,900 m^3).

Spain has an SPR with an approximate size of 120,000,000 barrels (19,000,000 m^3).

Sweden has an SPR with an approximate size of 13,290,000 barrels (2,113,000 m^3).

The United Kingdom has created a strategic reserve, the size is unknown.

Russia

Russia has begun accumulating strategic reserves of refined products which will be held by Rosneftegaz, the state-owned company. The reserves will be held at commercial refineries, Transneft facilities, and state reserve facilities. The current planned size is 14,665,982 barrels (2,331,704.8 m^3).

Switzerland

Switzerland has SPRs consisting of gas, diesel, jet fuel and heating oil for 4.5 months of consumption. The reserves were created in the

1940s and were used for the first time in 2005 following Hurricane Katrina.

Middle East

Iran: In April 2006 the Fars News Agency reported that Iran has begun plans to create an SPR. The National Iranian Oil Company (NIOC) has begun construction of 15 crude oil storage tanks with a planned capacity of 10,000,000 barrels (1,600,000 m^3). In August 2008, Iran announced plans to expand their SPR with a new facility on Kharg Island containing 4 tanks holding 1,000,000 barrels (160,000 m^3) each. Iran's SPR facilities are:

- Ahwaz. 4 storage tanks, total capacity 2 million barrels (320,000 m^3).
- Omidiyeh. 3 storage tanks, total capacity 3 million barrels (480,000 m^3).
- Goureh. 6 storage tanks, total capacity 4 million barrels (640,000 m^3).
- Sirri Island. 1 storage tank, total capacity 500,000 barrels (79,000 m^3).
- Bahregansar. 1 storage tank, total capacity 500,000 barrels (79,000 m^3).
- Kharg Island. 4 storage tanks, total capacity 4 million barrels (640,000 m^3). Planned facility, not operational yet.

Kuwait: Kuwait has a joint stockpile held in South Korea. The deal gives South Korea first rights to purchase the oil. The current size of the stockpile is 2 million barrels (320,000 m^3).

Others: As of 1975 Israel is believed to have a strategic oil reserve equal to 270 days of consumption.

Jordan has strategic oil reserves equal to 60 days of consumption or 6,240,000 barrels (992,000 m^3).

North America

United States: The United States has the largest reported Strategic Petroleum Reserve with a total capacity of 727,000,000 barrels (115,600,000 m^3). If completely filled, the US SPR could theoretically replace about 60 days of oil imports as the US is estimated to import approximately 12,000,000 barrels per day (1,900,000 m^3/d) of crude oil. According to the US Department of Energy the facilities' maximum flow rate is limited to approximately 4,400,000 barrels per day (700,000 m^3/d) when filled to maximum, with flow rate declining as the reserve is drawn down. The U.S. facilities are in salt caverns with locations in:

- Bryan Mound - located near Freeport, Texas. Capacity of 226,000,000 barrels (35,900,000 m^3).
- Big Hill - located near Winnie, Texas. Capacity of 160,000,000 barrels (25,000,000 m^3).
- West Hackberry - located near Lake Charles, Louisiana. Capacity of 219,000,000 barrels (34,800,000 m^3).
- Bayou Choctaw - located near Baton Rouge, Louisiana. Capacity of 72,000,000 barrels (11,400,000 m^3).
- Richton, Mississippi. A planned facility.

The US has also organised the 2-million-barrel (320,000 m^3) Northeast Home Heating Oil Reserve to supply northeast homeowners during shortages.

California: The state of California is considering the creation of Strategic Fuels Reserve.

Oceania

New Zealand has a strategic reserve with a 2008 size of 170,000 tons or 1,200,000 barrels (190,000 m^3). Much of this reserve is based on ticketed option contracts with Australia, Japan, the United Kingdom and the Netherlands, which allow for guaranteed purchases of petroleum in the event of an emergency.

Hydrocarbon Exploration

Hydrocarbon exploration (or oil and gas exploration) is the search by petroleum geologists and geophysicists for hydrocarbon deposits beneath the Earth's surface, such as oil and natural gas. Oil and gas exploration are grouped under the science of petroleum geology.

Exploration Methods

Visible surface features such as oil seeps, natural gas seeps, pockmarks (underwater craters caused by escaping gas) provide basic evidence of hydrocarbon generation (be it shallow or deep in the Earth). However, most exploration depends on highly sophisticated technology to detect and determine the extent of these deposits using exploration geophysics. Areas thought to contain hydrocarbons are initially subjected to a gravity survey, magnetic survey, passive seismic or regional seismic reflection surveys to detect large scale features of the sub-surface geology. Features of interest (known as *leads*) are subjected to more detailed seismic surveys which work on the principle of the time it takes for reflected sound waves to travel through matter (rock) of varying densities and using the process of depth conversion to create a profile of the substructure. Finally, when a prospect has been identified and evaluated and passes the oil company's selection criteria, an exploration well is drilled in an attempt to conclusively determine the presence or absence of oil or gas.

Oil exploration is an expensive, high-risk operation. Offshore and remote area exploration is generally only undertaken by very large corporations or national governments. Typical shallow shelf oil wells (e.g. North Sea) cost US$10 – 30 million, while deep water wells can cost up to US$100 million plus. Hundreds of smaller companies search for onshore hydrocarbon deposits worldwide, with some wells costing as little as US$100,000.

Elements of a Petroleum Prospect

A prospect is a potential trap which geologists believe may contain hydrocarbons. A significant amount of geological, structural and seismic investigation must first be completed to redefine the potential hydrocarbon drill location from a lead to a prospect. Five geological factors have to be present for a prospect to work and if any of them fail neither oil nor gas will be present.

- A source rock - When organic-rich rock such as oil shale or coal is subjected to high pressure and temperature over an extended period of time, hydrocarbons form.
- Migration - The hydrocarbons are expelled from source rock by three density-related mechanisms: the newly-matured hydrocarbons are less dense than their precursors, which causes overpressure; the hydrocarbons are lighter medium, and so migrate upwards due to buoyancy, and the fluids expand as further burial causes increased heating. Most hydrocarbons migrate to the surface as oil seeps, but some will get trapped.
- Trap - The hydrocarbons are buoyant and have to be trapped within a structural (e.g. Anticline, fault block) or stratigraphic trap
- Seal or cap rock - The hydrocarbon trap has to be covered by an impermeable rock known as a seal or cap-rock in order to prevent hydrocarbons escaping to the surface
- Reservoir - The hydrocarbons are contained in a reservoir rock. This is a porous sandstone or limestone. The oil collects in the pores within the rock. The reservoir must also be permeable so that the hydrocarbons will flow to surface during production.

Exploration Risk

Hydrocarbon exploration is a high risk investment and risk assessment is paramount for successful exploration portfolio management. Exploration risk is a difficult concept and is usually defined by assigning confidence to the presence of five imperative geological factors, as discussed above. This confidence is based on data and/or models and is usually mapped on Common Risk Segment Maps (CRS Maps). High confidence in the presence of imperative geological factors is usually coloured green and low confidence coloured red. Therefore these maps are also called Traffic Light Maps, while the full procedure is often referred to as Play Fairway Analysis. The aim of such procedures is to force the geologist to objectively assess all different geological factors. Furthermore it results in simple maps that can be understood by non-geologists and managers to base exploration decisions on.

Terms Used in Petroleum Evaluation

- Bright spot - On a seismic section, coda that have high amplitudes due to a formation containing hydrocarbons.
- Chance of success - An estimate of the chance of all the elements within a prospect working, described as a probability. High risk prospects have a less than 10% chance of working, medium risk prospects 10-20%, low risk prospects over 20%. Typically about 40% of wells recently drilled find commercial hydrocarbons.
- Dry hole - A formation that contains brine instead of oil.
- Flat spot - An oil-water contact on a seismic section; flat due to gravity.
- Hydrocarbon in place - amount of hydrocarbon likely to be contained in the prospect. This is calculated using the volumetric equation - GRV x N/G x Porosity x Sh x FVF
- GRV - Gross rock volume - amount of rock in the trap above the hydrocarbon water contact
- N/G - net/gross ratio - proportion of the GRV formed by the reservoir rock (range is 0 to 1)
- Porosity - percentage of the net reservoir rock occupied by pores (typically 5-35%)
- Sh - hydrocarbon saturation - some of the pore space is filled with water - this must be discounted
- FVF - formation volume factor - oil shrinks and gas expands when brought to the surface. The FVF converts volumes at reservoir conditions (high pressure and high temperature) to storage and sale conditions
- Lead - a structure which may contain hydrocarbons

- Play - A particular combination of reservoir, seal, source and trap associated with proven hydrocarbon accumulations
- Prospect - a lead which has been fully evaluated and is ready to drill
- Recoverable hydrocarbons - amount of hydrocarbon likely to be recovered during production. This is typically 10-50% in an oil field and 50-80% in a gas field.

Licensing

Petroleum resources are typically owned by the government of the host country. In the USA most onshore (land) oil and gas rights (OGM) are owned by private individuals. Sometimes this is not the same person who owns the surface rights. In this case oil companies must negotiate terms for a lease of these rights with the individual who owns the OGM. In most nations the government issues licences to explore, develop and produce its oil and gas resources, which are typically administered by the oil ministry. There are several different types of licence. Typically oil companies operate in joint ventures to spread the risk, one of the companies in the partnership is designated the operator who actually supervises the work.

- Tax and Royalty - Companies would pay a royalty on any oil produced, together with a profits tax (which can have expenditure offset against it). In some cases there are also various bonuses and ground rents (license fees) payable to the government - for example a signature bonus payable at the start of the licence. Licences are awarded in competitive bid rounds on the basis of either the size of the work programme (number of wells, seismic etc.) or size of the signature bonus.
- Production Sharing contract (PSA) - A PSA is more complex than a Tax/Royalty system - The companies bid on the percentage of the production that the host government receives (this may be variable with the oil price), There is often also participation by the Government owned National Oil Company (NOC). There are also various bonuses to be paid. Development expenditure is offset against production revenue.
- Service contract - This is when an oil company acts as a contractor for the host government, being paid to produce the hydrocarbons.

Bibliography

Anthony H.: *The Gulf and the West: Strategic Relations and Military Realities.* Boulder, Colo.: Westview Press, 1988.

Armistead, George: *Safety in Petroleum Refining and Related Industries*, New York: John G. Simmons & Co., 1950.

Black, Brian: *Petrolia: The Landscape of America's First Oil Boom.* Baltimore: Johns Hopkins University Press, 2000.

Bradford, Samuel A.: *Practical Handbook of Corrosion Control in Soils.* Edmonton, Canada: CASTI Pub., 2000.

Chowdhary, L.R. : *Petroleum Geology of the Cambay Basin, Gujarat, India*, Indian Petroleum, Delhi, 2004.

Crelling, J.C. : *Applied Coal Petrology: the Role of Petrology in Coal Utilization*, Academic Press, New York, 2008.

Diana Davids: *Oil in Texas: The Gusher Age, 1895–1945.* Austin: University of Texas Press, 2002.

Diwan, Parag and Ashutosh Karnatak: *Handbook of Natural Gas Technology and Business*, Pentagon Energy Press, Delhi, 2009.

Edward W.: *United States Oil Policy and Diplomacy: A Twentieth-Century Overview.* West-port, Conn.: Greenwood Press, 1983.

Engler, Robert: *The Politics of Oil: A Study of Private Power and Democratic Directions.* New York, Macmillan, 1961.

Folk, R.L.: *The Petrology of Sedimentary Rocks*, Hemphill Publishing Co., Austin, 2001.

Garverick, Linda: *Corrosion in the Petrochemical Industry.* Materials Park, Ohio: ASM International, 2000.

Gray, Earle: *The Great Canadian Oil Patch*, Maclean-Hunter, Ltd., Toronto, 1970.

Günter Simader: *Fuel Cells and Their Applications.* New York: VCH, 1996.

Hartshorn, J. E.: *Oil Companies and Governments: An Account of the International Oil Industry in Its Political Environment.* London: Faber and Faber, 1962.

John J.: *Sketches in Crude-oil*, J. Horace McFarland Company, Harrisburg, 1896.

Kennedy, John. L.: *Oil and Gas Pipeline Fundamentals*, Tulsa, PennWell Books, 1993.

Malhotra, Raman : *Mines and Minerals Laws*, Global India, Delhi, 2011.

Marsh, Steve: *Anglo-American Relations and Cold War Oil: Crisis in Iran.* New York: Palgrave MacMillan, 2003.

Miller, J.P. : *Fluvial Processes in Geomorphology*, Freeman, San Francisco, CA, 1964.

Mohr, Anton. *The Oil War*. Harcourt, Brace, 1926. Hyperion Press, 1976.

Norman J.: *Nontechnical Guide to Petroleum Geology, Exploration, Drilling, and Production*. PennWell Corporation, Delhi, 2001.

O'Connor, Harvey *The Empire of Oil*. Monthly Review Press, 1955

Peter F.: *The Problems of Plenty: Energy Policy and International Politics.* Berkeley: Uni-versity of California Press, 1985.

Philip, George: *Oil and Politics in Latin America: Nationalist Movements and State Companies*. Cambridge University Press, Cambridge, 1982.

Rathi, Rakesh : *Petroleum Refining Processes*, SBS Pub, Delhi, 2007.

Rister, Carl Coke: *Oil! Titan of the Southwest*. Norman: University of Oklahoma Press, 1949.

Robert J.: *Oil and the Middle East War: Europe in the Energy Crisis.* Cambridge: Center for International Affairs, 1976.

Schneider, Steven A.: *The Oil Price Revolution*. Baltimore, 1983.

Schweitzer, Philip A.: *Handbook of Corrosion Resistant Piping*. Malabar, FL: R.E. Krieger, 1985.

Sidney Russell Cooke: *The Oil Trusts and Anglo-American Relations*. Macmillan, 1924

Srivastava, Rajesh K. , Ch. Sivaji and N.V. Chalapathi Rao: *Indian Dykes: Geochemistry Geophysics and Geochronology*, Narosa, Delhi, 2008.

Thomas, L.: *Handbook of Practical Coal Geology*: John Wiley & Sons, Chichester, 1992.

Tiwari, S.K. : *Ore Geology, Economic Minerals And Mineral Economics*, Atlantic Publishers, Delhi, 2010.

Vervalin, Charles H.: *Fire Protection Manual for Hydrocarbon Processing Plants*. Houston: Gulf Publishing Co, 1985.

Walter G.: *Photometry and Polarization in Remote Sensing*, New York, 1985.

Watkins, Eric: *The Middle Eastern Environment*, London, The British Society for Middle Eastern Studie, 1995.

William L. Leffler: *Petrochemicals in Nontechnical Language*. Tulsa, Okla.: PennWell, 2001.

Index

❑❑❑